MASONRY: RESEARCH, APPLICATION, AND PROBLEMS

A symposium sponsored by
ASTM Committees C-7 on Lime,
C-12 on Mortars for Unit Masonry,
and C-15 on Manufactured Masonry Units
Bal Harbour, FL, 6 Dec. 1983

ASTM SPECIAL TECHNICAL PUBLICATION 871

John C. Grogan, Brick Institute of America
Region Nine, and
John T. Conway, Santee Cement Company,
editors

ASTM Publication Code Number (PCN)
04-871000-07

1916 Race Street, Philadelphia, PA 19103

Library of Congress Cataloging in Publication Data

Masonry, research, application, and problems.
(ASTM special technical publication; 871)
"ASTM publication code number (PCN) 04-871000-07."
Includes bibliographies and index.
1. Masonry—Congresses. I. Grogan, John C.
II. Conway, John T. III. ASTM Committee C-7 on Lime.
IV. ASTM Committee C-12 on Mortars for Unit Masonry.
V. ASTM Committee C-15 on Manufactured Masonry Units.
VI. Series.
TA670.M384 1985 693′.1 85-3852
ISBN 0-8031-0402-2

Library of Congress Catalog Card Number: 85-3852

NOTE

The Society is not responsible, as a body, for the statements and opinions advanced in this publication.

Printed in Baltimore, MD
April 1985

Dedication

This symposium is dedicated to the honor of J. I. Davison and to his high ideals of sponsoring masonry symposia to stimulate the sharing of theoretical and practical research and also to further the spirit of cooperation and unity within the masonry industry.

Foreword

The symposium on Masonry: Research, Application, and Problems was held in Bal Harbour, FL on 6 Dec. 1983. ASTM Committees C-7 on Lime, C-12 on Mortars for Unit Masonry, and C-15 on Manufactured Masonry Units sponsored the meeting. John T. Conway, Santee Cement Company, and John C. Grogan, Brick Institute of America Region Nine, presided as symposium cochairmen, and are editors of the publication. J. Gregg Borchelt, Masonry Institute of Houston-Galveston, Russell H. Brown, Clemson University, Kenneth A. Gutschick, National Lime Association, Alan H. Yorkdale, Brick Institute of America, and Lewis J. Yost, consultant, served as the symposium committee.

Related ASTM Publications

Masonry: Materials, Properties, and Performance, STP 778 (1982), 04-778000-07

Masonry: Past and Present, STP 589 (1975), 04-589000-07

A Note of Appreciation to Reviewers

The quality of the papers that appear in this publication reflects not only the obvious efforts of the authors but also the unheralded, though essential, work of the reviewers. On behalf of ASTM we acknowledge with appreciation their dedication to high professional standards and their sacrifice of time and effort.

ASTM Committee on Publications

ASTM Editorial Staff

Contents

Introduction

The masonry symposium, sponsored by ASTM Committee C-7 on Lime, C-12 on Mortars, and C-15 on Manufactured Masonry Units, was established in 1974 by J. Ivan Davison to whose memory this fourth symposium is dedicated. The symposium, as intended by Mr. Davison, is a continuing activity of the ASTM committees interested in masonry and provides a very important opportunity for the exchange of ideas and information to benefit the entire masonry industry.

The first symposium, held in June of 1974, served as a base for future symposia and as such included a historical review of masonry and its components as well as papers on masonry research and new developments in masonry. *Masonry: Past and Present, ASTM STP 589*, published in August of 1975, resulted from this first symposium. In June of 1976 the second symposium was held. It was similar in scope to the first symposium but contained more papers on the then current testing and research programs. Although no publication resulted from this symposium, several of the papers did appear in the *Journal of Testing and Evaluation*.

Masonry: Materials, Properties, and Performance, ASTM STP 778, published in September of 1982, was the result of the third masonry symposium that was held in December of 1980. This symposium provided a forum for the then current research on masonry units, mortar and grout and their components, and masonry assemblages. The majority of the papers dealt with the performance of masonry assemblages.

This publication is the result of the fourth masonry symposium, *Masonry: Research, Application, and Problems*, held in December of 1983. The title of this symposium was chosen by the symposium committee in an effort to encourage the submission of papers dealing with field application and end use problems with masonry as well as masonry research. The Call for Papers was intended to reach those interested in all aspects of masonry in the United States as well as several foreign countries. The committee was successful in this effort as illustrated by the papers contained herein. In addition to papers on research, dealing primarily with masonry assemblages, also included are papers on test methods and field application as well as field problems. These papers are authored by some of the industry's most knowledgeable people, and the information they have shared should be useful in understanding many of the problems that are experienced with the application of masonry in the field. But more importantly, this information can be used in future masonry design and construction to avoid problems. These papers will also serve to indicate some areas in which more information is needed, and hopefully, will encourage other investigators to address these areas.

A great deal has been learned about masonry construction in the thousands of years since man began stacking rocks to provide shelter. Very impressive gains in masonry engineering and masonry material properties have been made, but new techniques, ideas, and problems perpetually flow from the minds and the experience of people involved with masonry, universities active in masonry research, the material manufacturers, engineers, architects, builders, and others. Future symposia will continue to provide a means of exchange of information on these new techniques and ideas and solutions to the problems, and will provide a valuable source of information about masonry construction. The end result, the application of this knowledge, will assure more durable and economical masonry construction.

John T. Conway

Santee Cement Company, Holly Hill, SC 29059; symposium cochairmen and editor.

John C. Grogan

Brick Institute of America Region Nine, 8601 Dunwoody Place, Suite 507, Atlanta, GA 30338; symposium cochairman and editor.

Albert W. Isberner[1]

Composition of Hardened Masonry Mortar: A Critical Review of Test Methods and Calculations

REFERENCE: Isberner, A. W., **"Composition of Hardened Masonry Mortar: A Critical Review of Test Methods and Calculations,"** *Masonry: Research, Application, and Problems, ASTM STP 871*, J. C. Grogan and J. T. Conway, Eds., American Society for Testing and Materials, Philadelphia, 1985, pp. 3–14.

ABSTRACT: Masonry mortar samples removed from masonry evidencing distress may be petrographically and chemically analyzed for composition. The success in reconstructing the original mortar composition depends upon the method of analysis, the availability of the parent materials or proper assumptions regarding their composition, and considerations during calculation. This report highlights various considerations for any analysis of mortar composition.

KEY WORDS: masonry mortar, composition, data analysis

During the preparation of ASTM Methods for Preconstruction and Construction Evaluation of Mortars for Plain and Reinforced Unit Masonry (C 780), an array of test methods was considered for inclusion within the standard. The specific tests of ASTM C 780 were selected as being the most direct toward identification of specific mortar properties and mortar composition. Further, the test methods were included to allow isolation of a single physical property for more critical examination.

In addition to those methods within ASTM C 780, consideration was given toward methods that would allow assessing the compatibility of masonry mortar with masonry units. These compatibility specimens were conceived as masonry assemblages either built during construction or removed from an existing structure. With the passage of time and, more importantly, the activities of ASTM Committee E-6, test methods are now available for assessing the compatibility of masonry materials and masonry units.

[1] Consulting engineer, 4213 Commercial Way, Glenview, IL 60025.

Past, but persisting, interest in chemical analyses leading to the composition of hardened mortars was deemphasized with a preference toward quality control testing of mortars and inspection during actual construction. Testing of hardened mortars for composition remains a topic of interest, mostly because quality control of mortars during construction is inadequate, lacking, or nonexistent. The water content and cement to aggregate ratio tests within ASTM C 780 coupled with sporadic spot checks of batching sequence and quantities appeared much easier and more beneficial than to remove a sample of the hardened mortar and subject it to wet chemical analysis. Quality control exercised during construction did, and still does, appear to be more beneficial than post-construction analysis.

Masonry Mortars

While the majority of persons interested in masonry are aware of ASTM Specification for Mortar for Unit Masonry (C 270), our immediate interest toward establishing the composition of hardened mortars, which were to have been prepared in accordance with this specification, justifies their review. Simply, ASTM C 270 brings together individual products covered by other ASTM product standards, such as ASTM C 150, C 207, C 91, and C 144. Some, but not all, of these product specifications incorporate chemical requirements for the individual product.

Masonry Materials

The composition of individual masonry mortar components, which may comprise a part of an ASTM C 270 mortar, are reported in Table 1. The

TABLE 1—*Composition of masonry mortar ingredients.*

OXIDE	PORTLAND CEMENT	PORTLAND BLAST FURNACE SLAG CEMENT	MASONRY CEMENT	LIME: HIGH CALCIUM (QUICK)	LIME: DOLOMITIC	LIME: HYDRAULIC (QUICK)	FINE AGGREGATE: CLAY	FINE AGGREGATE: SILICEOUS	FINE AGGREGATE: LIMESTONE DOLOMITIC	COLORING COMPOUND
CAO	62.90	40.90	58.90	96.80	42.35	74.00	0.93	1.88	29.45	17.20
SIO2	21.70	33.90	11.40	1.00	0	21.70	62.36	97.02	.14	2.50
AL2O3	4.40	9.60	2.70	0.60	0	1.80	17.72	.69	.04	.35
FE2O3	3.40	.78	1.67	0.20	0	0.60	3.10	.68	0.10	60.00
MGO	3.60	11.40	.92	0	30.44	0.70	2.53	.01	21.12	1.04
SO3	2.50	3.08	1.41	.02	0	0	0	.03	0	.89
LOSS ON IGN	0	0	20.60	1.80	27.21	0	0	.01	0	15.90
INS RES	0	0	1.13	0	0	0	0	0	0	0
NA2O	.57*	0.70*	.02	0	0	0	0	.05	.01	0
K2O	0	0	.08	0	0	0	0	.05	.01	0
REFERENCE	1	1	5	2	3	3	4	5	3	5

Footnotes and References
0 No value reported in reference.
* Reported value includes K2O.
References, as listed in bibliography.

listing, as referenced, should be considered realizing that the composition of each material will vary slightly within a production plant and more significantly between plants.

Portland Cement—The oxide content and analysis of portland cement is reasonably fixed. However, the oxide content will vary within and between individual brands and types of cement.

Lime—The term "lime" may reflect a high calcium lime, a dolomitic lime, or a hydraulic lime. The oxide content of these three distinct "limes" varies significantly.

Masonry Cement—The composition of a masonry cement will vary with materials available to the cement manufacturer. Plasticizing materials selected include those capable of modifying the workability characteristics of mortars produced therewith, while maintaining adequate strength. Because of economics, those plasticizing materials most readily available to the cement manufacturer are often qualified and used as the plasticizing components of the masonry cement.

Sand—Masonry sand may be either natural or manufactured fine aggregate. These classifications suggest pure natural silica or manufactured limestone (calcareous or dolomitic) fine aggregates. Combinations of siliceous and calcareous fine aggregate occur in nature so neither type of aggregate should be considered pure. The proportion of calcium and silica in the aggregate will vary depending upon the source.

Admixtures—Although acceptance with approval is stipulated within specification ASTM C 270, the admixtures of immediate concern would be those inorganic admixtures that alter the chemical composition of the mortar and, consequently, the subsequent calculations during mortar analysis. At present, admixtures such as clay-shale, clay, colloid silica, and blast-furnace slag might be encountered within a blind, hardened masonry mortar sample.

Masonry Mortars

By intent, the present and past ASTM C 270 mortar specifications recognize mortars either prepared using prescribed mix proportions or prepared using a laboratory designed combination of cementitious material and fine aggregate, which laboratory tests demonstrate attainment of specified minimum cube compressive strength and water retention. The Proportion Specifications of ASTM C 270 limit the fine aggregate content to 2¼ to 3 times the sum of the cementitious materials. The Property Specifications of ASTM C 270, historically, allowed the aggregate content to be 2¼ to 3½ times the cementitious volumes.

Sampling for Hardened Mortar Tests

The success in breaking down and analyzing a hardened mortar sample depends upon the proper sampling of the hardened mortar and the sampling of the parent materials, that is, cement, lime, masonry cement, sand, and admixtures. If the parent materials are not available, bill of ladings or delivery tickets should be sought so the approximate mortar composition can be established. Additionally, the sensibility of the calculated proportions should be appraised by a person knowledgeable as to masonry construction practices.

Tests of Hardened Mortars

Analysis Considerations—The debate as to how much information should be supplied to the analyst will continue. Having the right answer before the analysis may cause the person completing the test to stop when in range. Conversely, the attainment of the wrong answer may cause additional verification of the data. The intended mortar type should be known by the chemist, in this writer's opinion, unless the test method is under study.

A successful analysis will depend more upon the ability of the sampler to get representative samples of the hardened mortar and parent materials than upon the chemistry. Selective oxide analysis can be very precise and accurate; assumptions during calculations can significantly influence the reported values. Regardless of the calculated mix proportions, the numbers must reflect the quality of a workable mortar as used or be discarded.

ASTM C 85—The present ASTM Test Method for Cement Content of Hardened Portland Cement Concrete (C 85-66, reapproved 1973), would at first appear to be the method to select for compositional analysis of hardened masonry mortar. The method would rapidly negate any immediate interest because of the precautionary statements within the scope of the method. The scope states:

> "This method of determining the cement content of concrete is applicable to hardened portland cement concretes except those containing certain aggregates or combinations of aggregates or admixtures which yield significant amounts of dissolved calcium oxide (CaO) and dissolved silica (SiO_2) under the conditions of the test."

The test method is depicted in the flow chart in Fig. 1. After determining the quantities of the oxides present, the calculations involve assuming of a CaO content of cement equal to 63.50% and a SiO_2 content of 21.00%. Properly, the note recommends, "Whenever possible, the known values should be used." The assumed oxide, either CaO or SiO_2, yielding the lowest calculated cement content is used for subsequent calculations. It should be noted also that the fine aggregate is measured for acid-soluble SiO_2 or CaO during this test.

The person using this method for masonry mortar should be aware that portland cement-lime masonry mortars and masonry cement mortars may, and generally do, contain CaO in excess of 63.5%, so the calculations are

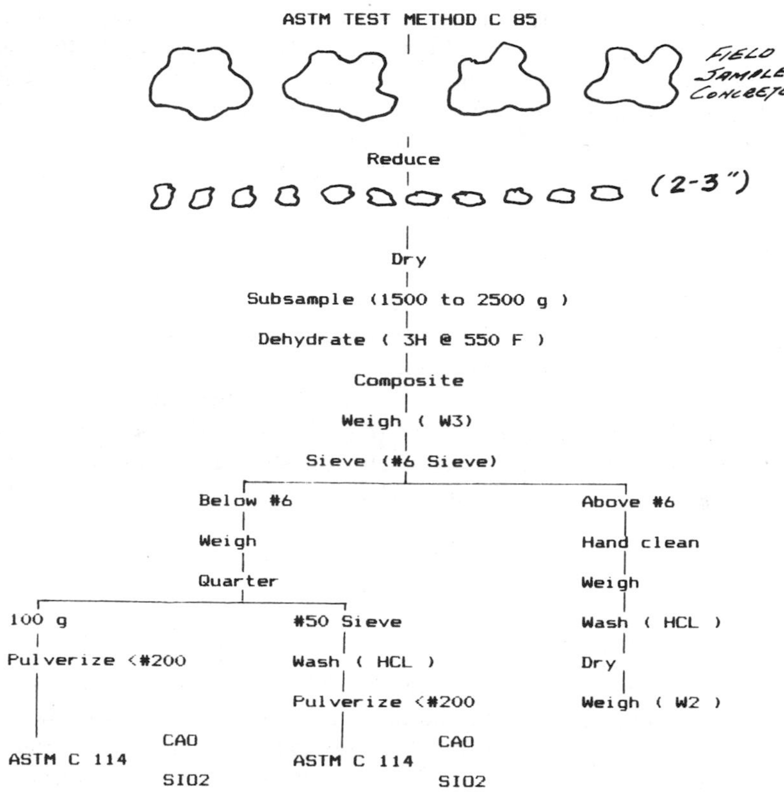

FIG. 1—*Chemical analysis of concrete.*

forced toward the SiO_2 oxide determination. It is interesting to note that the reciprocal of the assumed SiO_2 value is 4.76, whereas the reciprocal of the assumed CaO value is 1.57. The SiO_2 determination needs to be 3.03 times as precise as the CaO determination for equality to prevail.

Knowing that ASTM C 85 discourages the use where aggregates or combinations of aggregates or admixtures may interfere, other oxides are often considered for the base calculation of cement content. MgO or SO_3 or FeO might be chosen. Again, the more an analyst knows about the parent material, the better the compositional calculation can be.

Maleic Acid Method—An alternate method to ASTM Test Method for Cement Content of Hardened Portland Cement Concrete (C 85) was developed by Tabikh et al. [*6*] to allow measuring the cement content of concrete. This test method involves sample preconditioning and weighings followed by partial digestion using an alcoholic solution of maleic acid. Incremental weight readings during preconditioning and after acid digestion allow expressing the cement content of the concrete.

The method was recognized as being desirable, as the composition of the parent construction materials need not be known or available. Interferences recognized by the developers of the method and others include fly ash and fly ash combined with portland cement hydration. Although the aggregates tested by the method developer were not identified, the fact that aggregates are not affected by the acid treatment suggests that the carbonated aggregates and, consequently, carbonated portland cement hydration products are residue.

The direct applicability to masonry mortar should be questioned by anyone anticipating using the method, as the carbonated portland cement hydration products (presumably eventually 100%) and the carbonated hydrated calcium component of dolomitic lime would not be dissolved. Although the fact that carbonation of masonry mortars is well documented, the rate and even the extent of carbonation, as a function of time, needs to be seriously considered before the test data can be judged sensible. Masonry cements containing fine ground limestone pose a similar, yet more complicated, problem for the analyst.

Petrography

Through petrographic analysis the examination of the hardened mortar may provide additional information as to the composition of the masonry mortar. ASTM Recommended Practice for Petrographic Examination of Hardened Concrete (C 856) provides an excellent reference for procedures which may be applied to the examination of masonry mortar. The petrographer knowledgeable in concrete technology will have to compensate for the basic differences between concrete and masonry mortars. As is indicated in Section 2 of the Scope, the included procedures are considered "...applicable to the examination of samples of all types of hardened hydraulic-cement mixtures, including ... mortar, grout, plaster, stucco, terrazzo and the like." Regardless of the emphasis toward concrete and concrete construction, a petrographer investigating a masonry mortar problem should utilize this recommended practice. Additionally, the person requesting the text or interpreting and applying the petrographic report should be aware of the contents of the recommended practice.

The petrographer should work independently during the early portion of any mortar analysis leading to calculations of original composition. As the chemist needs to know something about mortar composition before selecting a precise analytical procedure, the petrographic report should be supplied to the chemist prior to chemical analysis so interferences during oxide analyses are minimized. Additionally, the petrographer's report is necessary for sensible calculated compositions.

Masonry Mortars—Calculated Compositions

General Considerations

Before any calculations commence, it should be recognized that masonry mortar preparation at the construction site is by either shovelsful or volumetric measure. The precise sand content cannot always be attained because of sand bulking, long days, etc., so any calculations leading to a mortar type identification should first consider the ratio of the cementitious materials singly. For example, a mortar prepared using one bag of cement and one (23-kg [50-lb]) bag of lime is a Type N mortar regardless of sand content. If the aggregate ratio is outside of the range (2¼ to 3 times cementitious volume), the mortar should be classified as a Type N mortar and further identified as being either oversanded or undersanded. It is considered proper to precede such classifications by stating that the mortar does not conform of ASTM C 270.

To highlight the confusion that may prevail during a mortar type selection based upon oxide contents, the data in Table 2 reflects the CaO and SiO_2 content of portland cement-lime mortars of ASTM C 270 Types M, S, N, and O. All calculations are based on assumptions incorporated within the table. With perfect wet chemistry and a CaO content of 14.3 for a mortar sample, the mortar type can be readily established as a Type N mortar. Referring to Fig. 2, which represents a plot of the data in Table 2, the mortar type can now be readily identified as a Type S, N, or O mortar.

When the calculations commence after the oxide analysis has been completed, the data in Table 3 are normally generated. Commencing with the oxide analysis, the mix proportions are eventually obtained while using the same assumptions incorporated within Table 2. Concern regarding the chemical analysis of an "as-received" sample, as is indicated in Table 3, is unwarranted as the proportions of oxides in mortars remain the same, relative one to the other, regardless of moisture and carbonation content. It is improper to use these data in conjunction with Table 2 and Fig. 2. The resulting mortar would be classed by the writer as a Type O and a Type N mortar, both being oversanded.

Specific Considerations

The specific entries and associated calculations reflected in Tables 2 through 4 are detailed in Table 5. As listed, the parent materials, excepting sand, were not available so oxides were assumed. Sand reportedly from the original source was received, chemically analyzed, and used to perfect the calculated mortar composition.

If either the parent materials or a manufacturer's certification of chemical

TABLE 2—*Oxide analysis of ASTM C 270 mortars: calculated values.*

MORTAR TYPE	M	M	S	S	N	N	O	O	K*	K*
MIX PROPORTI										
CEMENT, POR	1	1	1	1	1	1	1	1	1	1
CEMENT, MAS										
LIME, HYDRA	.25	.25	.5	.5	1.25	1.25	2.5	2.5	4	4
SAND, MASON	2.8125	3.75	3.3775	4.5	5.0625	6.75	7.875	10.5	11.25	15
WATER										
MIX, PROPORT										
CEMENT, POR	94	94	94	94	94	94	94	94	94	94
CEMENT, MAS	0	0	0	0	0	0	0	0	0	0
LIME, HYDRA	10	10	20	20	50	50	100	100	160	160
SAND, MASON	225	300	270.2	360	405	540	630	840	900	1200
WATER	0	0	0	0	0	0	0	0	0	0
ADMIXTURES	0	0	0	0	0	0	0	0	0	0
TOTAL WT. OF	329	404	384.2	474	549	684	824	1034	1154	1454
MIX PROPORTI										
CEMENT, POR	28.57	23.27	24.47	19.83	17.12	13.74	11.41	9.09	8.15	6.46
CEMENT, MAS	0	0	0	0	0	0	0	0	0	0
LIME, HYDRA	3.04	2.48	5.21	4.22	9.11	7.31	12.14	9.67	13.86	11.00
SAND, MASON	68.39	74.26	70.33	75.95	73.77	78.95	76.46	81.24	77.99	82.53
TOTAL, % DRY	100.	100.	100.	100.	100.	100.	100.	100.	100.	100.
CAO IN PORT	.63	.63	.63	.63	.63	.63	.63	.63	.63	.63
SIO2 IN POR	.21	.21	.21	.21	.21	.21	.21	.21	.21	.21
CAO IN MASO	.315	.315	.315	.315	.315	.315	.315	.315	.315	.315
SIO2 IN MAS	.105	.105	.105	.105	.105	.105	.105	.105	.105	.105
CAO IN LIME	.4235	.4235	.4235	.4235	.4235	.4235	.4235	.4235	.4235	.4235
MGO IN LIME	.3044	.3044	.3044	.3044	.3044	.3044	.3044	.3044	.3044	.3044
SIO2 IN LIM	0	0	0	0	0	0	0	0	0	0
CAO IN SD	0	0	0	0	0	0	0	0	0	0
SIO2 IN SD	0	0	0	0	0	0	0	0	0	0
INS RES IN	100	100	100	100	100	100	100	100	100	100
CAO % OF DR	19.29	15.71	17.62	14.28	14.64	11.75	12.33	9.82	11.00	8.73
SIO2 % OF DR	6.00	4.89	5.14	4.16	3.60	2.89	2.40	1.91	1.71	1.36
OTHER, % OF	74.71	79.41	77.24	81.55	81.76	85.36	85.28	88.27	87.29	89.91

Footnotes

* ASTM C 270 mortar type K included as they will be encountered for next decade.

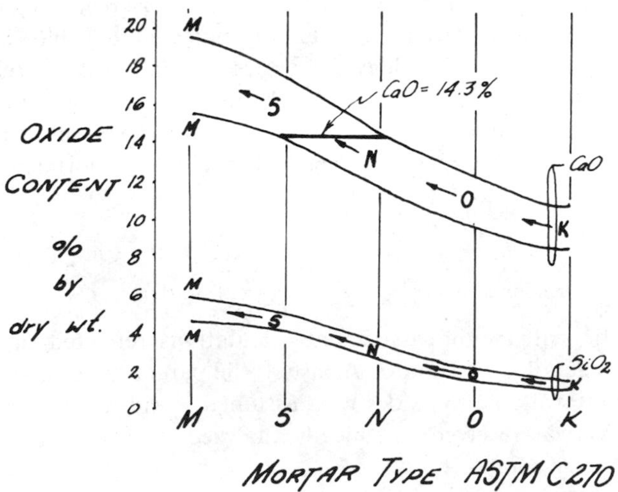

FIG. 2—*Calculated oxide contents of ASTM C 270 mortars.*

TABLE 3—*Calculation of mortar composition using chemical analysis data.*

R OW	ANI MORTAR ANALYSIS		
2	CAO	12.98	11.80
3	SIO2	1.66	1.73
4	MGO	3.83	3.67
5	SO3		
6	INS. RES.	65.56	67.96
7	LOI 105 C	0.47	0.36
8	550 C	3.70	3.89
9	950 C	9.78	8.69
10	OTHER	2.02	1.90
11	TOTAL	100.00	100.00
12	OXIDE ANALYSIS - CEM		
13	CAO	63.50	63.50
14	SIO2	21.00	21.00
15	MGO	3.50	3.50
16	SO3	3.50	3.50
17	OXIDE ANALYSIS LIM		
18	CAO	42.35	42.35
19	MGO	30.44	30.44
20	SIO2	0.00	0.00
21	OXIDE ANALYSIS- SAND		
22	CAO	6.96	6.96
23	MGO	2.13	2.13
24	SIO2	0.20	0.20
25	INS RES	80.60	80.60
26	CORRECTIONS - SAND		
27	INS. RES.	81.34	84.32
28	CAO	5.66	5.87
29	MGO	1.73	1.80
30	SIO2	0.16	0.17
31	CORRECTIONS TO OXIDE		
32	CAO	7.32	5.93
33	SIO2	1.50	1.56
34	MGO	2.10	1.87
35	SO3	0.00	0.00
36	INS. RES.	81.34	84.32
37	LOI 105 C	0.47	0.36
38	550 C	3.70	3.89
39	950 C	9.78	8.69
40	OTHER	2.02	1.90
41	TOTAL	108.22	108.52
42	DIFF/CHECK	8.22	8.52
43	CEMENT CONTENT		
44	CAO BASE	11.62	9.42
45	SIO2 BASE	7.13	7.44
46	LIME CONTENT (SIO2 B		
47	PC CAO	4.53	4.72
48	CAO IN LIME	2.79	1.21
49	CAO IN LIME HYDRATE	6.59	2.86
50	SUMMARY		
51	PC	7.13	7.44
52	HL	6.59	2.86
53	SD	81.34	84.32
54	TOTAL	95.06	94.61
55	MORTAR PROPORTIONS,		
56	PC	1	1
57	HL	0.92	0.38
58	SD	11.41	11.34
59	TOTAL WEIGHT	13.33	12.72
60	% PC	7.50	7.86
61	% HL	6.93	3.02
62	% SD	85.57	89.12
63	MORTAR PROPORTIONS B		
64	PC	1	1
65	HL	2.17	0.90
66	SD	13.40	13.33

TABLE 4—*Effect of assumed oxide value on mortar composition results.*

OXIDE ANALY	A	B	C	D	E
CAO	13.02	13.02	13.02	13.02	13.02
SIO2	3.61	3.61	3.61	3.61	3.61
MGO	3.83	3.83	3.83	3.83	3.83
SO3					
INS. RES.	65.56	65.56	65.56	65.56	65.56
LOI 105 C	.47	.47	.47	.47	.47
550 C	3.7	3.7	3.7	3.7	3.7
950 C	9.78	9.78	9.78	9.78	9.78
OTHER	.03	.03	.03	.03	.03
TOTAL	100	100	100	100	100
OXIDE ANALY					
CAO	63.5	62.5	64.5	63.5	63.5
SIO2	21	21	21	20	22
MGO	3.5	3.5	3.5	3.5	3.5
SO3	3.5	3.5	3.5	3.5	3.5
OXIDE ANALY					
CAO	0	0	0	0	0
SIO2	0	0	0	0	0
MGO	0	0	0	0	0
SO3	0	0	0	0	0
OXIDE ANALY					
CAO	42.35	42.35	42.35	42.35	42.35
MGO	30.44	30.44	30.44	30.44	30.44
SIO2	0	0	0	0	0
OXIDE ANALY					
CAO	0	0	0	0	0
MGO	0	0	0	0	0
SIO2	0	0	0	0	0
INS RES	100	100	100	100	100
CORRECTIONS					
INS. RES.	65.56	65.56	65.56	65.56	65.56
CAO	0	0	0	0	0
MGO	0	0	0	0	0
SIO2	0	0	0	0	0
CORRECTIONS					
CAO	13.02	13.02	13.02	13.02	13.02
SIO2	3.61	3.61	3.61	3.61	3.61
MGO	3.83	3.83	3.83	3.83	3.83
SO3	0	0	0	0	0
INS. RES.	65.56	65.56	65.56	65.56	65.56
LOI 105 C	.47	.47	.47	.47	.47
550 C	3.7	3.7	3.7	3.7	3.7
950 C	9.78	9.78	9.78	9.78	9.78
OTHER	.03	.03	.03	.03	.03
TOTAL	100	100	100	100	100
DIFF/CHECK	0	0	0	0	0
CEMENT CONT					
CAO BASE	20.50	20.83	20.19	20.50	20.50
SIO2 BASE	17.19	17.19	17.19	18.05	16.41
LIME CONTEN					
PC CAO	10.92	10.74	11.09	11.46	10.42
CAO IN LIM	2.10	2.28	1.93	1.56	2.60
CAO IN LIM	4.97	5.37	4.56	3.68	6.14
SUMMARY					
PC	17.19	17.19	17.19	18.05	16.41
HL	4.97	5.37	4.56	3.68	6.14
SD	65.56	65.56	65.56	65.56	65.56
TOTAL	87.72	88.12	87.31	87.29	88.11
MORTAR PROP					
PC	1	1	1	1	1
HL	0.29	0.31	0.27	0.20	0.37
SD	3.81	3.81	3.81	3.63	4.00
TOTAL WEIG	5.10	5.13	5.08	4.84	5.37
% PC	19.60	19.51	19.69	20.68	18.62
% HL	5.66	6.10	5.23	4.22	6.97
% SD	74.74	74.39	75.09	75.11	74.41
MORTAR PROP					
PC	1	1	1	1	1
HL	0.68	0.73	0.62	0.48	0.88
SD	4.48	4.48	4.48	4.27	4.69

TABLE 5—*Mortar composition calculation format. (Notes: R stands for Row; . . . stands for "include all values.")*

ROW	MORTAR ANALYSIS	VALUE	SOURCE OR FORMULA
2	CAO	12.98	Test Result
3	SIO2	1.66	do
4	MGO	3.83	do
5	SO3	0	Not Tested
6	INS RES	65.56	Test Result
7	LOI 105C	0.47	do
8	550C	3.70	do
9	950C	9.78	do
10	OTHER	2.02	100.00 Sum(R2...R9)
11	TOTAL	100.00	Constant
12	OXIDE ANALYSIS OF PORTLAND CEMENT		
13	CAO	63.50	Assumed
14	SIO2	21.00	do
15	MGO	3.50	do
16	SO3	3.50	do
17	OXIDE ANALYSIS OF HYDRATED LIME		
18	CAO	42.35	Assumed
19	MGO	30.44	do
20	SIO2	0.00	do
21	OXIDE ANALYSIS OF SAND (PARENT)		
22	CAO	6.96	Test Result
23	MGO	2.13	do
24	SIO2	0.20	do
25	INS RES	80.60	do
26	CORRECTIONS FOR SAND		
27	INS RES	81.34	100*R6/R25
28	CAO	5.66	R27*R22/100
29	MGO	1.73	R27*R23/100
30	SIO2	0.16	R27*R24/100
31	CORRECTIONS TO OXIDE ANALYSIS		
32	CAO	7.32	R2-R28
33	SIO2	1.50	R3-R30
34	MGO	2.10	R4-R29
35	SO3	0.00	R5
36	INS RES	81.34	R27
37	LOI 105C	0.47	R7
38	550C	3.70	R8
39	950C	9.78	R9
40	OTHER	**2.02**	**R10**
41	TOTAL	**108.22**	**Sum (R32...R40)**
42	**DIFF/CHECK**	**8.22**	**R41-100.00**
43	**CEMENT CONTENT CALCULATION**		
44	**CAO BASE**	11.62	**R32/R13/100**
45	**SIO2 BASE**	7.13	**R33/R14/100**
46	LIME CONTENT (SIO2 governs)		
47	PC CAO	4.53	R45/R13/100
48	CAO IN LIME	2.79	R32-R47
49	CAO IN LIME HYDRATE	6.59	(R48/R18)*100
50	SUMMARY		
51	PC	7.13	R45
52	HL	6.59	R49
53	SD	81.34	R36
54	TOTAL	95.06	Sum (R51...R53)
55	MORTAR PROPORTIONS, BY WT		
56	PC	1	R51/R51
57	HL	0.92	R52/R51
58	SD	11.41	R53/R51
59	TOTAL WEIGHT	13.33	Sum (R56...R58)
60	%PC	7.50	(R56/R59)*100
61	%HL	6.93	(R57/R59)*100
62	%SD	85.57	(R58/R59)*100
63	MORTAR PROPORTIONS, BY VOL		
64	PC	1	R56/(94/94)
65	HL	2.17	R57/(94/40)
66	**SD**	**13.40**	R58/(94/80)

composition are available, tests if necessary should be completed and the more accurate values should replace the assumed values.

The combined petrographic and chemical analyses will allow other modifications to the computer program and the calculated results. If a masonry cement is encountered or if aggregates or admixtures containing soluble calcium or silica are involved, the calculations without the parent materials being analyzed are exceedingly complex and may lead to erroneous results.

Using the spread sheet capabilities to a greater degree, it is possible to rapidly show the effect of assumed values. The data of Table 4 show the effect of increasing and then decreasing the assumed CaO and SiO_2 content of the portland cement. As will be noted, the data show that assumed value does have a significant effect on the final, calculated mortar composition.

Summary and Conclusions

Computer-assisted manipulation of test results of hardened mortar from petrographic and chemical analysis yield the approximate composition of the hardened mortar. If the parent materials are not available, calculation may lead to an indeterminate solution to the composition.

References

[*1*] Roy, D. M. and Idorn, G. M., "Hydration, Structure, and Properties of Blast Furnace Slag Cements, Mortars, and Concrete," *American Concrete Institute Journal* No. 6, *Proceedings*, Vol. 79, Technical Paper 79-43, Nov./Dec. 1982, pp. 444–457.

[*2*] Schlitt, W. J. and Healy, G. W., "Characteristics of Lime: A Comparison and Scaling Down of the Coarse Grain Titration Test and the ASTM Slaking Rate Test," *The Reaction Parameters of Lime, ASTM STP 472*, American Society for Testing and Materials, Philadelphia, 1970, pp. 143–160.

[*3*] Boynton, R. S., *Chemistry and Technology of Limestone*, 2nd ed., Wiley, New York, 1980.

[*4*] Ries, H., *Clays, Their Occurrences, Properties, and Uses*, 2nd ed., Wiley, New York, 1914.

[*5*] Isberner, A. W., Glenview, IL, Data in file.

[*6*] Tabikh, A. A., et al, "A Method for the Determination of Cement Content in Concrete," draft copy, undated.

Edward Gazzola,[1] *Dino Bagnariol,*[1] *Janine Toneff,*[1] *and Robert G. Drysdale*[1]

Influence of Mortar Materials on the Flexural Tensile Bond Strength of Block and Brick Masonry

REFERENCE: Gazzola, E., Bagnariol, D., Toneff, J., and Drysdale, R. G., **"Influence of Mortar Materials on the Flexural Tensile Bond Strength of Block and Brick Masonry,"** *Masonry: Research, Application, and Problems, ASTM STP 871*, J. C. Grogan and J. T. Conway, Eds., American Society for Testing and Materials, Philadelphia, 1985, pp. 15–26.

ABSTRACT: This paper reports on 475 tests for the influence of mortar materials on the flexural tensile strength of brick and block masonry for tension normal to the bed joints. The bond wrench testing technique was used to test each joint of the prisms. The results indicated significant decreases in tensile bond for mortars made with portland cement and masonry cement instead of portland cement and lime. Tests using sand with a gradation near the middle of the gradation limits (sieved concrete sand) and the available masonry sand that has more than the maximum percent passing a 600-μm (No. 30) sieve showed that the masonry sand produced slightly better tensile bond. In addition, flexural tensile bond strengths from both laboratory- and field-prepared prisms were compared with results obtained by removing bricks from the top of a brick wall. The average strengths and the influence of variability are discussed in terms of behavior and in terms of possible design code approaches.

KEY WORDS: bed joints, block, bond, bond wrench, brick, concrete, field tests, flexure, lime, masonry, masonry cement, mortar, tensile strength, variability

The bond between mortar and masonry units for flexural tension normal to the bed joints is a topic where large differences in strengths and variabilities have been observed [*1–7*], and where allowable stresses differ significantly from country to country [*8, 9*]. The high variability of tensile bond strength within a sample has been recorded by many researchers. It seems to be an

[1] Master of engineering students, research engineer, and professor, respectively, Department of Civil Engineering and Engineering Mechanics, McMaster University, Hamilton, Ontario, Canada L8S 4L7.

unavoidable characteristic of this brittle form of failure where workmanship [*10*] and natural variations in the properties of the materials [*1,11,12,13*] lead to coefficients of variation in the order of 30 percent for laboratory conditions. The large differences caused by using different combinations of materials have indicated significant influences of mortar proportions [*1,4,5,12,14,15*], absorption and other properties of the units [*2,7,12,16*], age and curing [*17,18*], and workmanship [*10*].

While certain trends have been identified, the large number and range of variables have to date made it impossible to develop any quantifiable set of guidelines or specifications to assure consistent levels of tensile bond between mortars and masonry units. In general it has been shown that use of lime results in better bond than use of masonry cement [*1,2,5*]. Better bond is achieved with higher flows of the mortar [*1,4,5*], and lower initial rate of absorption (IRA) [*6,12,13*]. However, there are also many contradictory results. In design standards, the current approach is to assign specific allowable tensile bond stresses to combinations of mortars and masonry units which satisfy the appropriate material standards. However, this practice must either result in very low values [*8*] because of the large range of results, or allow higher values which have been to some extent calibrated to full-scale tests [*19*]. In the latter case, it has been shown that flexural tensile strength depends on the number of masonry units in a course [*17*] and the number of courses subjected to bending moments [*20*].

At the outset of this research project and during the course of preparing this paper, it has been apparent that a solution to the problem of extreme variability of tensile bond between mortar and masonry units will not be achieved easily. Therefore this paper provides some background information for Canadian materials and conditions. Also, from these results and similar published information, an attempt is made to propose a method of attacking this problem both in terms of planned research and development of design criteria.

Experimental Program

Until recently, tests for flexural bond strength with tension normal to the bed joints have usually been some form of beam test with various loading conditions. The common feature of these standard test procedures is that several mortar joints are subject to bending moments. Therefore the controlling critical combination of moment and strength implicitly provides a biased sample of joint strengths since all other joints in the specimen did not fail. The obvious solution is to test all joints. From research done using this approach [*20,21*] it is apparent that, at least from a research point of view, this is the logical process to follow. However, the natural extension of this is to consider introduction of a testing technique which could be efficiently used in both the field and the laboratory. Hughes and Zsembery [*18*] have designed a device called a bond wrench which is essentially a long cantilever beam with an attachment on one end for gripping individual masonry units. Because of the long cantilever, only modest weights need be used to provide sufficient bending moment to break a masonry unit off of the top of a wall or to successively break the bond between courses of units in field- or laboratory-prepared prisms. The feature of testing all joints and the suitability for both field and laboratory conditions led to adoption of this test technique in this research. Figure 1 includes a photograph of a brick prism being tested using the bond wrench.

Range of Tests

The first use of the bond wrench was in a preliminary field study where permission was given to "wrench" bricks off the top of a wall during construction. The mason also prepared prisms using the same mortar and brick. Table 1 contains the proportions and cube strength of the S_A mortar used on this site; Table 2 lists the initial rate of absorption (IRA) and compressive strength of the yellow brick (YB) used on this building. The flexural bond strengths for pulling bricks off the wall and for testing the field-produced prisms are shown in Series 1 to 3 in Table 3.

To gain further insight into the use of the bond wrench and other factors, the field mortar was reproduced in the laboratory and used with a combed finish red brick (RB1). Series 4 to 7 looked at the influence of the length of lever arm used for the bond wrench.

Series 8 used untooled joints and Series 9 used S_B mortar, which is the same as S_A mortar but with sieved concrete sand substituted for the regular masonry sand. Figure 2 contains the sieve analysis of the masonry sand shown as Curve C. Curve A shows the gradation of the sieved sand used in the concrete block test program; Curve B is the gradation of the sieved sand used for the clay brick test program.

Series 10 through 25 inclusive in Table 3 contain the results of a small parametric study where two bricks of slightly differing properties [seeTable 2 for brown brick (BB) and a second red brick (RB2)] were combined with four types each of S and N mortars. As shown in Table 1, the S_1 to S_4 mortars and N_1 to N_4 mortars contained different combinations of portland cement, lime, masonry cement, and masonry or sieved sand. All mortars had flows of approximately 120 percent. Only single types of portland cement, lime, and masonry cement were used.

Series 26 to 29 were flexural tensile bond tests using hollow 190 mm concrete blocks and the four types of S mortar mentioned above. These were autoclaved blocks and, as was the case for brick, all units were dry before use.

In designing the mortar mix, the proportions used were those specified in

FIG. 1—*Photograph of bond wrench.*

Canadian Standards [*9*]. The mortar sand used (Curve C) was typical for the Hamilton area whereas the sieved sand was concrete sand sieved to fit the middle range of gradation between the CSA limits.

Various flows were tried with the mason deciding that the 120 value gave about the best workability for all mortar mix designs. However, the workability was not thought to be equal. The masons preferred the mortars made with mortar sand and with masonry cement.

Discussion of Test Results

Field Construction—Series 1, 2, and 3 were for field construction tests. Comparison of Series 1 with Series 2, which were both done on the third day

TABLE 1—*Mortar.*[a]

Designation	Portland Cement	Masonry Cement	Lime	Sieved Sand[b]	Masonry Sand[b]	Cube Strength, N/mm^2
S_A	1.0	2.0 (1.45)	...	...	9.0 (8.25)	13.53
S_B	2.0	2.0 (1.45)	...	8.92 (8.25)	...	14.48
S_1	1.0	...	0.5 (0.2)	...	4.0 (4.81)	12.12
S_2	1.0	2.0 (1.45)	...	...	8.0 (9.58)	9.67
S_3	1.0	2.0 (1.45)	...	6.34 (8.0)	...	20.19
S_4	1.0	...	0.5 (0.2)	3.36 (4.24)	...	19.15
N_1	1.0	...	1.0 (0.39)	...	6.0 (7.22)	5.13
N_2	...	1.0	...	...	3.0 (4.96)	2.78
N_3	...	1.0	...	3.0 (5.20)	...	5.10
N_4	1.0	...	1.35 (0.54)	5.68 (7.16)	...	9.05

[a] Values are given as volumes and bracketed terms are weights, which were used for better quality control. All weights are relative weights with respect to the portland cement (PC) [or masonry cement (MC) where applicable].

[b] Sand proportions were adjusted to give better workability where possible.

TABLE 2—*Properties of masonry units.*

Unit	Symbol	*IRA*, $kg/m^2/min$	Compressive Strength, N/mm^2
Red brick (combed)	RB1	1.95	128.7
Yellow brick	YB	0.54	95.8
Red brick (smooth)	RB2	0.28	118.4
Brown brick	BB	0.33	118.9
Autoclaved block	BL	...	19.2

after laying, indicates a significant difference in tensile bond strength between bricks on the wall and bricks in prisms. Hughes and Zsembery [*18*] also noted that bond of bricks on the wall was lower but equated this difference to curing conditions. In this case curing was similar. Therefore there seems to be some indication that field construction of prisms may be better quality than regular construction. The Series 3 field prism tests at 28 days were quite simi-

TABLE 3—*Flexural tensile bond strength results.*

Series	Brick	Mortar	Age, days	Number of Tests	Mean, N/mm^2	COV^a, %	Remarks
1	YB	S_A	2.7	15	0.29	22.6	on wall
2	YB	S_A	3	20	0.40	19.4	fieldprisms
3	YB	S_A	28	26	0.41	20.8	fieldprisms
4	RB1	S_A	28	13	0.34	30.4	regular lever
5	RB1	S_A	28	14	0.30	27.1	1198 mm
6	RB1	S_A	28	16	0.34	27.1	588 mm lever
7	RB1	S_A	28	14	0.36	29.7	298 mm lever
8	RB1	S_A	28	23	0.37	37.3	untooled joints
9	RB1	S_B	28	21	0.35	41.1	
10	BB	S_1	28	15	0.97	20.8	
11	BB	S_2	28	15	0.69	9.8	
12	BB	S_3	28	15	0.60	18.2	
13	BB	S_4	28	15	0.55	30.6	
14	BB	N_1	28	15	0.94	20.3	
15	BB	N_2	28	15	0.72	14.8	
16	BB	N_3	28	15	0.55	15.7	
17	BB	N_4	28	15	0.74	23.7	
18	RB2	S_1	28	15	0.63	14.8	
19	RB2	S_2	28	15	0.65	18.7	
20	RB2	S_3	28	15	0.42	23.7	
21	RB2	S_4	28	15	0.59	10.4	
22	RB2	N_1	28	15	0.82	26.4	
23	RB2	N_2	28	15	0.62	34.3	
24	RB2	N_3	28	15	0.55	17.0	
25	RB2	N_4	28	15	0.59	26.8	
26	BLOCK	S_1	28	15	0.41	18.5	1.2 m lever
27	BLOCK	S_2	28	20	0.13	24.6	1.2 m lever
28	BLOCK	S_3	28	15	0.17	27.2	1.2 m lever
29	BLOCK	S_4	28	23	0.31	28.9	1.2 m lever

[a]COV is coefficient of variation.

lar to the 3-day Series 2 prism tests. It should be noted that the initial three days of curing were in a dry warm outside climate under sunny conditions. The remaining 25 days were in the laboratory at approximately 20°C and 50 percent relative humidity.

Aside from the comparison of prism versus wall data, these three series gave the lowest tensile bond results found to that point in brick masonry research at McMaster University. It also was the first field work and the first time that masonry cement had been used.

Influence of Lever Arm of Bond *Wrench and Other Factors*—Series 4 to 7 used a laboratory reproduction of the field mortar (S_A) and a red brick (RB1) which had been used in previous research [*14,15*]. The variable in these series was the length of the lever arm for the bond wrench. In varying the length from the standard 1838 mm down to 298 mm from the center of the brick, no significant difference was observed for flexural tensile bond strength. (Higher strain gradients are associated with the higher axial loads required to produce

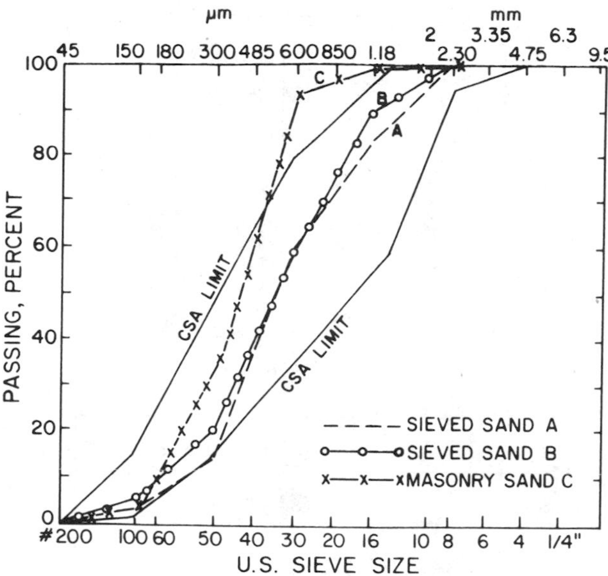

FIG. 2—*Gradation of mortar sands.*

equal bending moments for shorter lever arms. It was thought that this might have had some influence on this tensile strength.) The flexural tensile bond strengths for these tests were lower than for the field-produced specimens using the yellow brick (YB).

Series 8 was similar to the preceding four series except that the joints were not tooled. No influence could be detected. Series 9 substituted sieved sand (Curve A) for the masonry sand of mortar S_A to produce mortar S_B. Again no significant difference was apparent.

Influence of Mortar Proportions—Series 10 to Series 25 show results of a small parametric study of the influence of mortar proportions. As can be seen from Figs. 3*a* and 3*b*, there is generally a significant benefit in using lime instead of masonry cement for mortars made with masonry sand (mortar S_1 versus S_2 and N_1 versus N_2). For sieved sand there was a slight benefit in using lime instead of masonry cement, but strengths with sieved sand mortars were generally less than those using similar proportions of masonry sand. The similarity between results for Types S and N mortars is interesting. It is also interesting to note that there is a fairly consistent difference in results for relatively small IRA differences between the two types of bricks used.

Figure 4 illustrates the influence of Type S mortar proportions on the flexural tensile bond between mortar and concrete blocks. These results, which comprise Series 26 to 29 in Table 3, illustrate the better bond associated with use of lime instead of masonry cement (mortars S_1 and S_4 versus S_2 and S_3). For mortars containing lime, the use of sieved sand instead of masonry sand

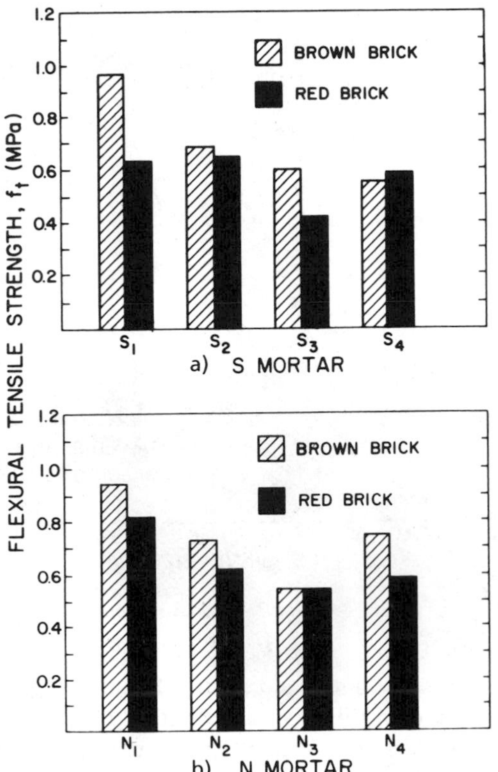

FIG. 3—*Influence of mortar materials for brick masonry.*

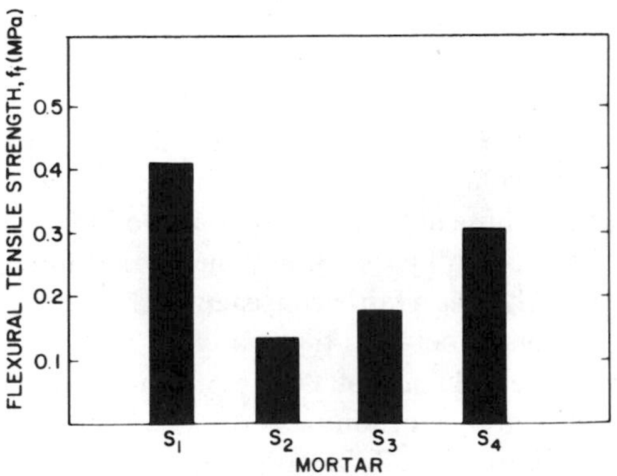

FIG. 4—*Influence of mortar materials for block masonry.*

resulted in a significant decrease in strength. The result was reversed with use of masonry cement.

The results of all these series are displayed in Fig. 5 to provide a combined illustration of the influence of the mortar proportions.

Influence of Initial Rate of Absorption—Figure 6 shows the results of the flexural bond tests for all the bricks used in this test program. Also shown are the results of previous beam tests [*13, 14*] which should give lower values than

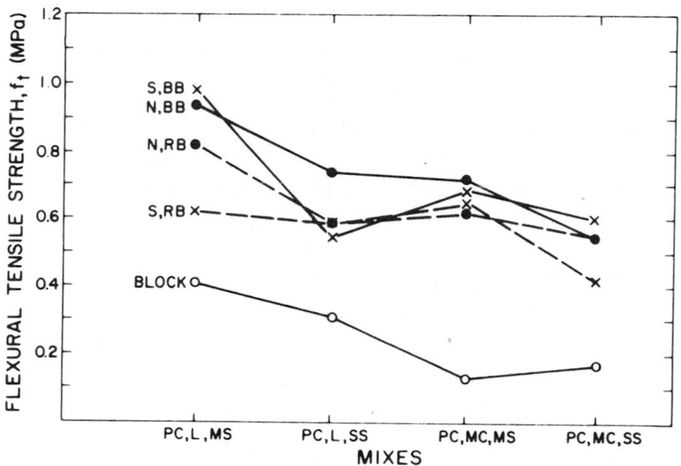

FIG. 5—*Summary of influence of mortar materials on flexural tensile bond.*

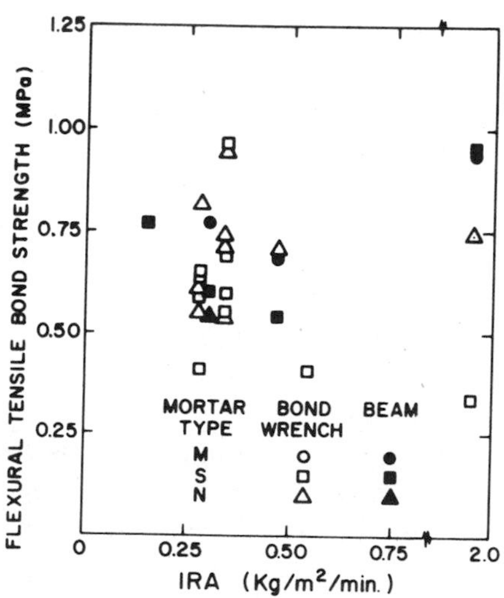

FIG. 6—*Relationship between bond and IRA.*

for bond wrench testing of all joints. As can be seen, no consistent trend is apparent.

Conclusions and Recommendations

1. The flexural tensile bond strengths reported in this paper and several of the referenced papers clearly illustrate the wide range of mean values which exist within the broad groups identified in design standards. The lower values of this range when judged in terms of characteristic strength (mean value minus 1.5 standard deviations) are of the same order of magnitude as currently specified allowable working stresses in North America. Therefore, it is suggested that some review of the unrestricted application of these allowable stresses is warranted.

2. The use of masonry cement has become widespread and in most parts of Canada is the predominant cementitious component. From conversations with masons, it seems that this trend has resulted to a large extent from the masons' performance and the resulting efficiency on the job. For example, the masons associated with this research much preferred the mortars with masonry cement and found it more difficult to work with the portland cement-lime mortars, especially those with coarse (sieved) sand. However, the use of masonry cement seems to have a direct correlation with significantly lower tensile bond strengths. Therefore it is suggested that the use of masonry cement should be reviewed.

3. There does not seem to be any correlation between tensile bond strength of mortar and initial rates of absorption of the masonry units. In general it does not seem that there is much chance of developing a simple universally applicable criteria for the individual materials to guarantee reasonably high levels of tensile bond. In fact it is suggested that a more meaningful approach would be to introduce qualification testing of the mortar-masonry unit combination. This would assure certain minimum levels of tensile bond.

4. The sand gradation limits are too wide and in particular include sand which is much too coarse for proper workability of the mortar. In addition it seems that sands that are somewhat finer than allowed will not adversely affect strengths and are generally preferable for better workmanship.

5. It is recommended that the bond wrench be adopted as the standard test for flexural tensile bond of mortar joints. In order that this can also be used in the field with relatively light loads, it is suggested that a standard lever arm in the order of 1.2 m or more be adopted.

6. As an alternative to mortar cube testing, it is suggested that field testing of tensile bond using the bond wrench be considered by standards writing bodies. In many situations tensile bond is much more important than compressive strength.

7. The statistical characteristics of flexural tensile strength of masonry walls composed of many masonry units per course and many courses are not

the same as those from tests of individual joints between masonry units. Therefore it suggested that additional work is necessary to correlate control tests of joints to wall strengths.

Acknowledgments

This research was carried out at McMaster University and was funded by Operating Grants from the Natural Sciences and Engineering Research Council of Canada and the Masonry Research Foundation of Canada. The authors appreciate the contribution of the mason's time made available through the Ontario Masonry Contractors Association and the Ontario Masonry Promotion Fund, and we also thank the Clay Brick Association of Canada and Domtar Inc., Clay Products Division for donating the bricks, and the Ontario Concrete Block Association for the donation of the blocks.

References

[*1*] Baker, L. R., "Some Factors Affecting the Bond Strength of Brickwork" in *Proceedings*, Fifth International Brick Masonry Conference, Washington, DC, Oct. 1979, pp. 62-72.

[*2*] Goodwin, J. F. and West, H. W. H., "A Review of the Literature on Brick/Mortar Bond" in *Proceedings*, British Ceramic Society—Load Bearing Brickwork, Vol. 7, No. 30, Sept. 1982.

[*3*] Isberner, A. W., "Properties of Masonry Cement Mortars," *Designing Engineering and Construction with Masonry Products*, Gulf Publishing Company, Houston, 1969.

[*4*] Jessop, E. L. and Langan, B. W., "Influence of Mortar Cube Strength Variability on the Measured Compressive and Flexural Strengths of Clay Masonry Prisms," *Proceedings, Fifth International Brick Masonry Conference*, Washington, DC, Oct. 1979, pp. 163-170.

[*5*] Neis, V. V. and Chow, D. Y. T., "Tensile Bond Testing of Structural Masonry Units" in *Proceedings*, Second Canadian Masonry Symposium, Ottawa, June 1980, pp. 381-395.

[*6*] West, H. W. H., "Current Masonry Research at the British Ceramic Research Association" in *Proceedings*, North American Masonry Conference, Boulder, CO, Aug. 1978, paper No. 6.

[*7*] Yorkdale, A. H., "Initial Rate of Absorption and Mortar Bond," *Masonry: Materials, Properties, and Performance, ASTM STP 778*, American Society for Testing and Materials, Philadelphia, 1982, pp. 91-98.

[*8*] Standards Association of Australia, "Australian Standard 1640-1974, SAA Brickwork Code," Sydney, N.S.W., 1974.

[*9*] Canadian Standards Association, "CAN3-S304-M78, Masonry Design and Construction for Buildings," Rexdale, Ontario, Canada, 1978.

[*10*] Matthys, J. E. and Grimm, C. T., "Flexural Strength of Nonreinforced Brick Masonry with Age" in *Proceedings*, Fifth International Brick Masonry Conference, Washington, DC, Oct. 1979, pp. 114-121.

[*11*] Baba, A., Kamimura, K. and Kato, S., "Design for Mix Proportion of Joint Mortar and Bond Strength" in *Proceedings*, Fifth International Brick Masonry Conference, Washington, DC, Oct. 1979, pp. 62-72.

[*12*] Satti, K. M. H. and Hendry, A. W., "The Modulus of Rupture of Brickwork," Third International Brick Masonry Conference, Essen, F.R.G., April 1973, pp. 155-160.

[*13*] Vanderkeyl, R., "Behavioural Characteristics of Brick Masonry," M.Eng. Thesis, McMaster University, Hamilton, Ontario, Canada, May 1979, 277 pages.

[*14*] Drysdale, R. G. and Hamid, A. A., "Effect of Eccentricity on the Compressive Strength of Brickwork" in *Proceedings*, British Ceramic Society, Load-Bearing Brickwork (7), No. 30, Stoke-on-Trent, United Kingdom, Sept. 1982, pp. 140-148.

[*15*] Drysdale, R. G. and Hamid, A. A., "Anisotropic Tensile Strength Characteristics of Brick Masonry" in *Proceedings*, Sixth International Brick Masonry Conference, Rome, Italy, May 1982, pp. 143–153.
[*16*] Drysdale, R. G. and Hamid, A. A., "Influence of the Characteristics of the Units on the Strength of Block Masonry" in *Proceedings*, Second North American Masonry Conference, College Park, MD, Aug. 1982.
[*17*] Baker, L. R. and Franken, G. L., "Variability Aspects of Flexural Strength of Brickwork" in *Proceedings*, Fourth International Brick Masonry Conference, Brugge, Belgium, April 1976.
[*18*] Hughes, D. M. and Zsembery, S., "A Method of Determining the Flexural Bond Strength of Brickwork at Right Angles to the Bed Joint" in *Proceedings*, Second Canadian Masonry Symposium, Ottawa, Canada, June 1980, pp. 73–86.
[*19*] Monk, C. B., "Transverse Strength of Masonry Walls," *Methods of Testing Building Construction, ASTM STP 166*, American Society for Testing and Materials, Philadelphia, 1954.
[*20*] Baker, L. R., "Measurement of the Flexural Bond Strength of Masonry" in *Proceedings*, Fifth International Brick Masonry Conference, Washington, DC, Oct. 1979, pp. 62–72.
[*21*] Brown, R. H. and Palm, B. D., "Flexural Strength of Brick Masonry Using the Bond Wrench" in *Proceedings*, Second North American Masonry Conference, College Park, MD, Aug. 1982.

DISCUSSION

John Melander[1] (*written discussion*)—How many different brands of masonry cements were used in the test paper? What were the physical properties of the various mortars tested; that is, compressive strength, according to ASTM Specification for Mortar for Unit Measuring (C 270), water retention, air content?

R. G. Drysdale (*author's closure*)—Only one of the commonly available brands of masonry cement was used in this research program. Table 1 lists the batch proportions and cube strengths for the various mortars used. The air content and water retention values are not available for the S_A and S_B mortars. For the other mortars they are as follows: $S_1 = 8.2$ and 68%, $S_2 =$ 12.0 and 60%, $S_3 = 12.0$ and 71%, $S_4 = 4.5$ and 76%, $N_1 = 7.5$ and 56%, $N_2 = 11$ and 64%, $N_3 = 12.0$ and 65%, and $N_4 = 6.4$ and 59%.

[1] Riverton Corporation, Riverton, VA 22651.

Albert W. Isberner[1]

A Test Method for Measuring the Efflorescence Potential of Masonry Mortars

REFERENCE: Isberner, A. W., **"A Test Method for Measuring the Efflorescence Potential of Masonry Mortars,"** *Masonry: Research, Application, and Problems, ASTM STP 871*, J. C. Grogan and J. T. Conway, Eds., American Society for Testing and Materials, Philadelphia, 1985, pp. 27-37.

ABSTRACT: Efflorescence of masonry construction is a natural occurrence when masonry is constructed in other than under normal conditions. Early age efflorescence may occur shortly after construction has been completed; late age efflorescence may occur if design and construction considerations and building maintenance are not properly attended. Efflorescence, once understood, is deemed beneficial toward identifying problems associated with building performance and indicating corrective action to be taken to remedy the situation.

An efflorescence test method for masonry mortars has been developed. The method is considered beneficial for isolating contributors toward efflorescence, be it adverse exposure conditions, cement retarders, or contaminated individual materials. The interpretation of test results allows isolation of contributing factors and remedial action. Practically, testing of masonry mortars will not guarantee that efflorescence will not appear on the surface of a structure.

KEY WORDS: alkali migration, efflorescence, masonry mortars, mortar composition, test method

During the past years in cement research, which involved both physical and chemical considerations, efflorescence of mixtures containing portland cement has proven to be a subject of primary interest. The interplay between physical and chemical characteristics of these mixtures coupled with the extreme variability that is common in the practical masonry construction industry and the in-service exposure of the buildings makes the subject of efflorescence of masonry even more interesting.

[1]Consulting engineer, 4213 Commercial Way, Glenview, IL 60025.

Classical reports regarding the subject of efflorescence have been available for years. The classic by Butterworth [1] remains the best overall reference regarding efflorescence of masonry. Others have approached the subject while being a bit more selective in the direction of their research programs. Surprisingly some more recent researchers seek the contributions made by a single construction material. Others fail to understand the basic physical-chemical factors involved.

Past Studies of Efflorescence

Over the years numerous test methods have been employed to allow demonstrating the physical-chemical characteristics of and necessary for the occurrence of efflorescence.

A simple wick test (Fig. 1) was first employed to cause a preferred migration of water from a source to an evaporative surface. The test method provides the preferred migration of water to an evaporative surface; the water reservoir, however, allows osmotic movement which influences total salts migrated to the evaporative surface. The test method proved simple to perform and showed that efflorescence will occur during the early periods after mortar preparation or will not occur. As will be indicated later, late age efflorescence shows that the efflorescence potential is ever present but remains dormant unless physically reactivated.

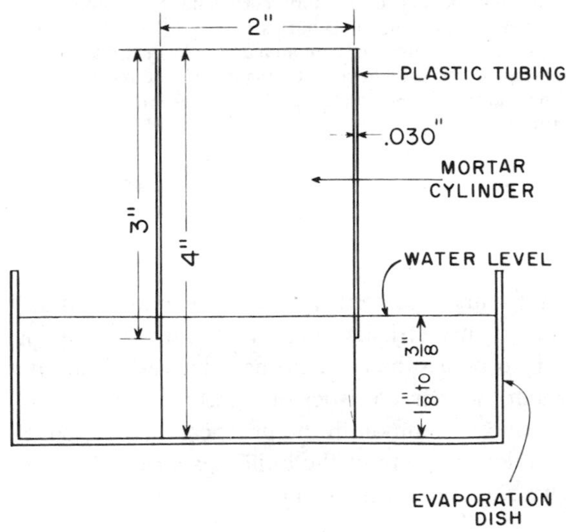

FIG. 1—*Wick efflorescence test specimen (courtesy of PCA).*

A more refined efflorescence test method (Fig. 2) was then developed to allow monitoring the water movements and the chemical migration within the mixture. A cylindrical specimen was provided with a central core reservoir during fabrication. During testing, the reservoir was maintained full of water under a slight hydrostatic head. Time for efflorescence to occur was subjectively visually appraised. The water inflow/outflow from the specimen was monitored to allow establishing water flow demand for efflorescence. At some appropriate time, the cylindrical test specimen was removed to a lathe where incremental samples were removed for wet chemical analysis. Mobility of the water-soluble alkalies was demonstrated.

To demonstrate the efflorescence potential of masonry as influenced by the combination of brick and mortar, an assemblage (Fig. 3) was equipped with a water reservoir to promote migration to the face of the masonry. This method was cooperatively tested by ASTM Subcommittee C12.02 during the period 1962 to 1964. The method's shortcomings appeared to be an inability to photographically duplicate the efflorescence rating chart and the between-laboratories variability. The method is still considered the best relative indicator of the efflorescence potential of masonry, albeit appraisal/rankings are subjective.

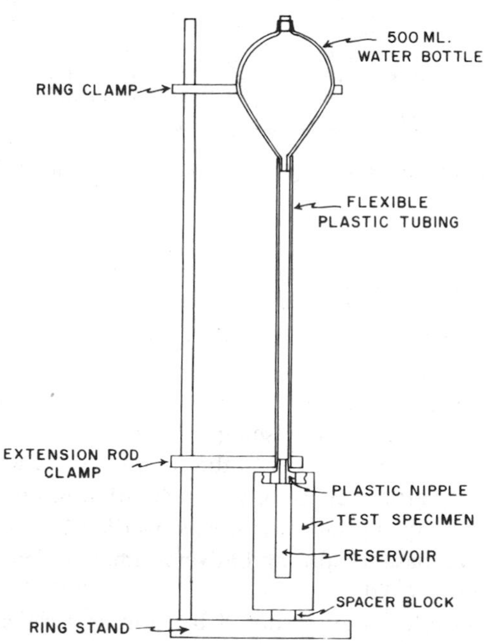

FIG. 2—*Cylinder efflorescence test specimen (courtesy of PCA).*

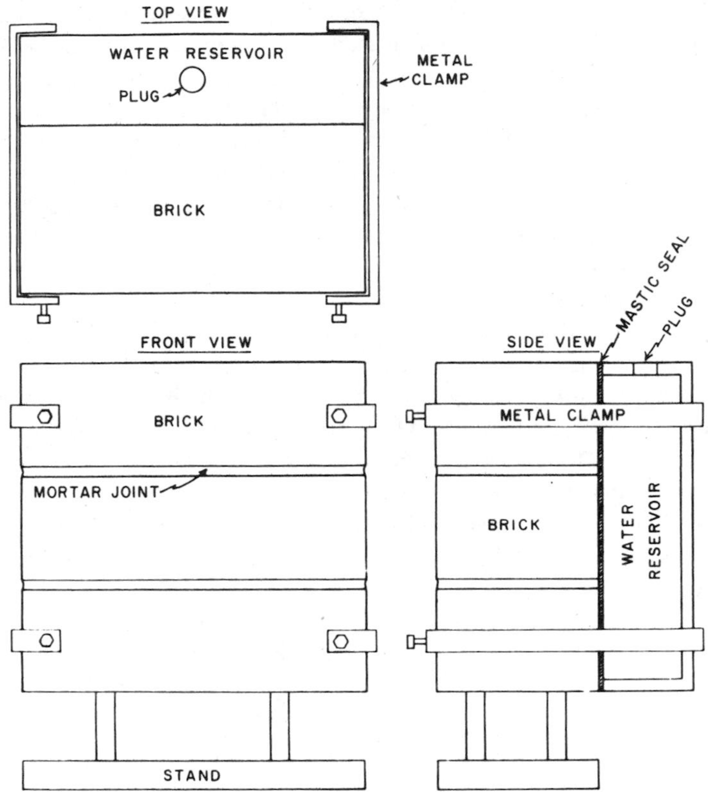

FIG. 3—*Masonry assemblage test specimen (courtesy of PCA).*

Present Studies of Efflorescence

Before embarking upon a new efflorescence test method, the practical aspects of masonry warrant discussion. All masonry construction materials may contribute to efflorescence. Under the right (wrong) conditions efflorescence will occur. An ASTM C67 non-efflorescing brick coupled with a masonry mortar in the proper environment will bring about an interplay wherein the brick may contribute toward efflorescence. Conversely, what has historically been a non-efflorescing masonry mortar may develop efflorescence when coupled with a low-absorption masonry unit and adverse ambient conditions, be these conditions too dry or too cold.

The occurrence of efflorescence should be used as a signal that all is not right either with the normal chemical reactions of the masonry materials or with the performance of the structure. As shown in Fig. 4, late age efflorescence highlights a problem area within a building. Efflorescence of greater

FIG. 4—*Late age efflorescence of Masonry.*

magnitude when detected may be reflecting climatic influences, incompatibility of construction materials, or poor workmanship, see Fig. 5. It is difficult in both the cases to visually examine a structure and efflorescence and then fault a specific material, construction practice, or workmanship.

The present method of test was developed to allow rapid isolation of possible

FIG. 5—*Early age efflorescence of masonry.*

contributors to the occurrence of efflorescence. The test apparatus is generally available in hardware stores; the test assemblage is sufficiently small to allow including temperature as a variable. Water migration into and from the test and control specimens are weighed; efflorescence is appraised visually using the companion control test specimen. The test method is available to interested readers from the author.

Test Method

The present test method involves fabricating both a control and an actual test specimen from a single paste or mortar sample. The specimen molds are used to contain the test sections during test. The control specimen mold is modified by installing a plug so evaporational forces, singly, affect the efflorescence; the test specimen mold is equipped with a water supply chamber to provide available water throughout the test section and at the evaporation surface. The basic plastic plumbing parts are depicted in Fig. 6. Figure 7 shows the method in use.

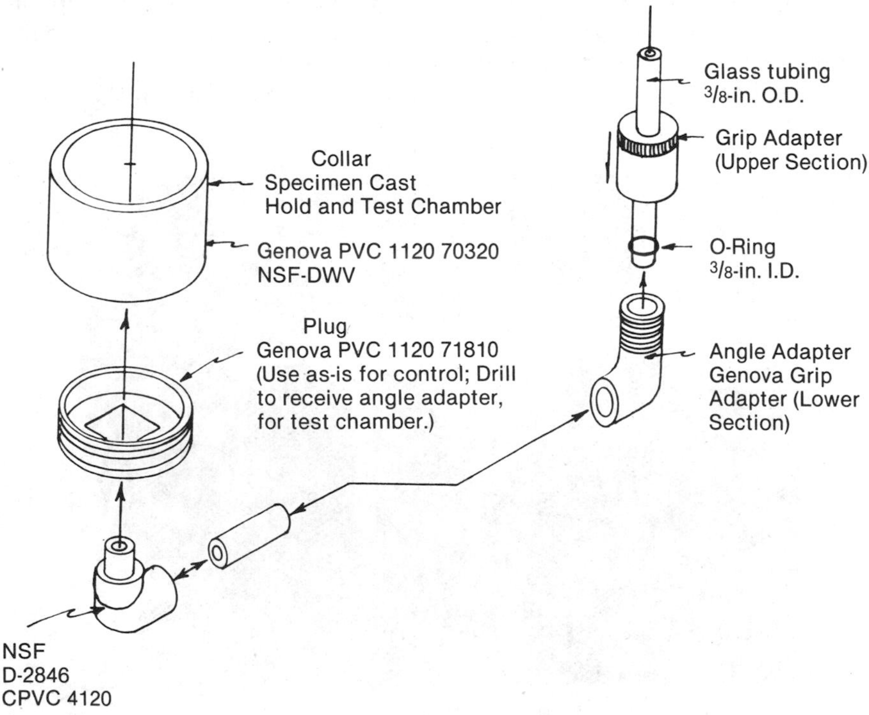

FIG. 6—*Plastic components for efflorescence test method.*

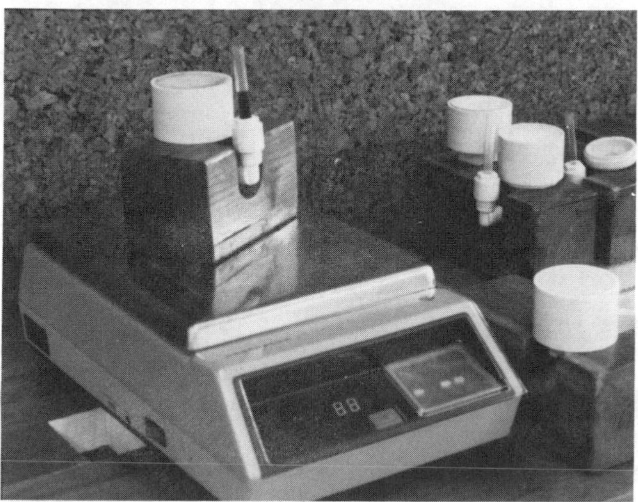

FIG. 7—*Efflorescence test specimens under test.*

The following past findings and observations will be of interest while using the test method.

1. Hydrostatic pressure is not a requirement. Evaporation forces have a greater influence on the occurrence of efflorescence than hydrostatic head.
2. Alkali migration and alkali metal migration is to be expected unless the alkalies and alkali metals are contained such that their water soluble characteristics are lessened. Blast furnace slags contain insoluble (water) alkali compounds.
3. Water addition and weight changes reflect interacting performance characteristics of the system. If weight changes exceed water additions, evaporation losses are high. If water additions cause an increase in weight, additional cement hydration is occurring. Carbonation of the mortar surface will allow weight increase with a release of water.
4. Efflorescence either will occur early after testing commences using the paste or mortar sections or will not occur, unless shrinkage cracking at the surface significantly reduces the water flow distance through the paste or mortar section.
5. Paste samples with high water to cement ratios can be rendered less prone to settlement if false setting of the cement is induced by either aerating the cement or heating the cement in an oven at 110°C for a period of 2 h.

Discussion of Test Results

Water Migration

The quantity of water flowing through a portland cement mixture is related to the water/cement ratio of the mixture. Unless the water/cement ratio is sufficiently high (generally considered to be above $W/C = 0.7$), water flow will diminish and abate as cement hydration occurs. Evaporation forces will cause further water flow through the mixture until the mixture attains internal conditions similar to ambient conditions. Temperature and humidity influence the water migration, as will any mortar ingredient which inhibits or retards portland cement hydration. Typical water loss curves are depicted in Fig. 8. The "chilled" and "heated" curves demonstrate the effect of temperature during the early period of hydration/storage on the performance of the test specimens.

Mortar Ingredients

In addition to the construction materials recognized by ASTM Specification for Mortar for Unit Masonry (C 270), a wide array of admixtures or alternatives are used during masonry construction. Inorganic admixtures such as clay, clay-shale, colloidal silica, coloring compounds, and organic admixtures are not rare. Testing of these admixtures too frequently involves entire structures and the performance of the mortar in the structure.

Testing

To allow isolating the wide array of variables and to allow appraising the combination as a mixture, the test method was developed. Under normal test-

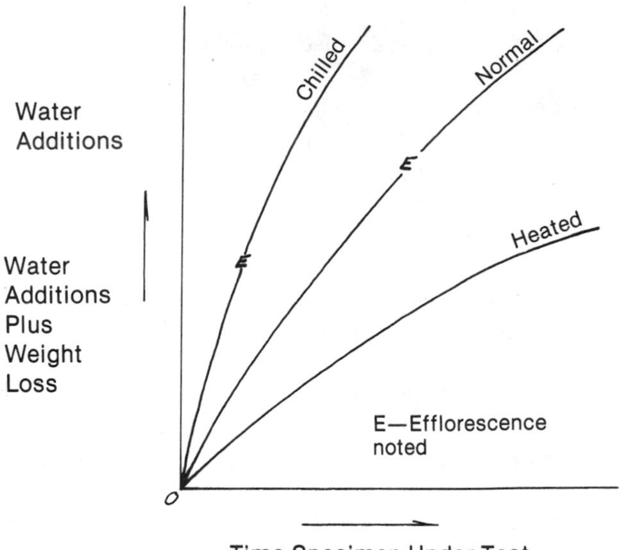

FIG. 8—*Efflorescence test data—water loss curves.*

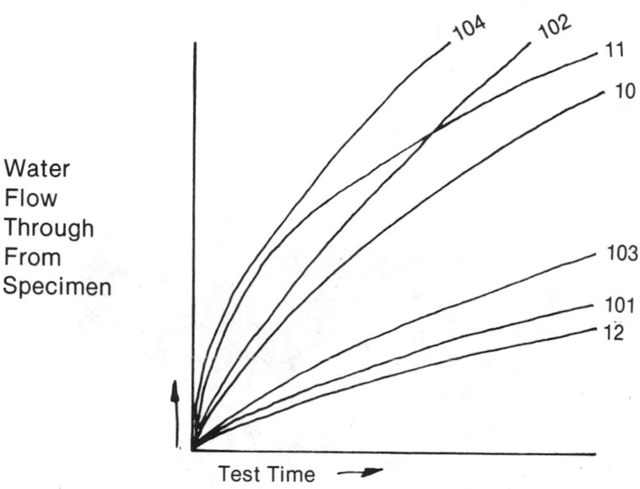

Example Data

Specimen	10	11	12	101	102	103	104
Masonry Cmt	1	1	1	1	1	1	1
Mason's Sand	3	3	3	3	3	3	3
Admix #1 Accel.	0	0	0	1	0	0	0
2 Inorgan.	0	0	0	0	1	0	0
3 Retard.	0	0	0	0	0	1	0
4 Color	0	0	0	0	0	0	1
W/C	N[a]	N	N	<N	>N	<N	>N
Consistency	N	N	N	N	N	N	N
Temp Pretest	N	<N	>N	N	N	N	N
Test	N	N	N	N	N	N	N
R.H. Pretest	N	N	N	N	N	N	N
Pretest	N	N	N	N	N	N	N
Eff. Pat. (Rank)[b]	3	5	0	1	4	2	6

[a]N Control mix value.
[b]Rank based on effloresence time and curve.

FIG. 9—*Efflorescence test data.*

ing, the mortar of use or the mortar intended to be used is prepared along with additional mortars without inclusion of specific ingredients. Testing then involves the introduction of other known influences, that is, temperature and humidity conditions. Typical data are shown in Fig. 9.

As will be noted, documentation of the basic mixture and physical characteristics of the mixtures is required and eases proper interpretation of the rest results. Plotting the water addition, be it water additions, singly, or water addition and weight loss data, reflects a part of permeability and evaporational forces. When efflorescence occurs, be it slight or heavy, the water addition or

FIG. 10—*Efflorescence specimen sampling for alkali migration.*

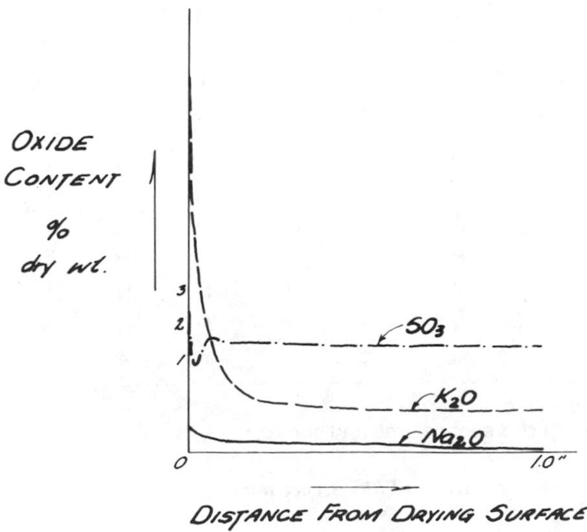

FIG. 11—*Alkali migration within paste specimen (courtesy of PCA).*

days till efflorescence can be used to identify and to rank varying mortars, thereby, isolating significant contributors toward efflorescence.

The method also lends itself to measuring the salts of efflorescence. After disassembling the test specimen apparatus, the collared test specimen can be placed in a lathe and cut to varying depths, Fig. 10. By monitoring the position of the cutting tool, the sample thickness, that is, the distance from the evaporative surface, can be established and later correlated with the oxide analysis of the sample, Fig. 11.

Summary

Although the writer is not convinced that a single test or multiple test of masonry mortar will allow selecting non-efflorescing masonry mortars in service, the test method described will allow establishing the contribution attributed to a single construction material or a combination of construction materials toward increasing/decreasing efflorescence.

Acknowledgment

The author wishes to thank the Portland Cement Association Research and Developmental Laboratories for permission to use specific photographs and figures. The author is also much appreciative of the chemistry knowledge imparted to him by Messrs. Nathan Greening and William G. Hime.

Reference

[*1*] Butterworth, B., "Efflorescence and Staining of Brickwork," *The Brick Bulletin*, Vol. 3, No. 5, The National Federation of Clay Industries, London, 1950, pp. 3-11.

John H. Matthys[1] *and Bunlert Chanprichar*[1]

Ultimate Strength Flexural Theory for Reinforced Brick Masonry

---

REFERENCE: Matthys, J. H. and Chanprichar, B., **"Ultimate Strength Flexural Theory for Reinforced Brick Masonry,"** *Masonry: Research, Application, and Problems, ASTM STP 871*, J. C. Grogan and J. T. Conway, Eds., American Society for Testing and Materials, Philadelphia, 1985, pp. 38–66.

ABSTRACT: The results of a two-phase experimental research program to develop an ultimate flexural strength theory for reinforced brick masonry is reported. Twenty beam-column specimens and corresponding prisms were tested to determine the flexural compressive stress distribution for high and low strength brick masonry constructed with Type S portland cement lime mortar with loading both parallel and perpendicular to the bed joints. This data established the parameters for defining an ultimate compressive stress block for brick masonry. Reinforced brick masonry beam tests (still in progress) have determined the applicability of the developed theory to reinforced flexural elements. Results show that (1) the ultimate compressive stress-strain relationship for brick masonry under flexural loading and axial loading are different; (2) there is a correlation between the ultimate flexural compressive strain and prism strength; and (3) an ultimate strength theory reasonably predicts the capacity of reinforced brick masonry beams.

KEY WORDS: ultimate strength theory, flexural design, beams, stress-strain curves, brick masonry

Nomenclature

a Distance from P to neutral axis
a_1 Distance from major thrust P_1 to neutral axis
a_2 Distance from minor thrust P_2 to neutral axis
b Width of member
c Distance from compression edge of member to neutral axis
C Resultant of internal compressive force in masonry
ϵ Strain
ϵ_c Strain on compression edge of member

[1]Associate professor civil engineering and director of Construction Research Center, and graduate student in civil engineering, respectively, University of Texas at Arlington, Arlington, TX 76019.

ϵ_x Masonry strain
f_c Extreme flexural compressive stress
f_0 Average compressive stress in masonry compression zone
$f'm$ Axial compressive strength of masonry
f_x Flexural masonry compressive stress
k_1, k_2, k_3 Coefficients related to magnitude and position of internal compressive force in masonry compression zone
M Bending moment
m_0 Modified moment term, M/bc^2
P Applied force
P_1 Major thrust
P_2 Minor thrust
x Distance from neutral axis to compression fiber

The current structural flexural design procedure for reinforced brick masonry in the U.S.A. would be classified as a working stress design procedure. Although ultimate flexural design theory has been widely accepted for reinforced concrete design, its application to reinforced masonry (clay or concrete) design in the U.S.A. does not exist. A literature survey indicates essentially no experimental efforts in this area in the U.S.A. Although there are some applications of this theory in some European countries and in Australia, the developed design calculation procedures were not based on actual measured brick masonry flexural stress distribution.

Objectives and Scope

The overall objective of this study was to investigate the behavior and performance of rectangular reinforced brick masonry members under ultimate load in flexure including: (1) generating the actual bending stress distribution and axial stress distribution; (2) observing strain variation across brick masonry members subjected to bending; (3) quantifying the stress-strain behavior for different directions of compressive stress with respect to the bed joints; (4) investigating the potential application of an ultimate strength design concept to reinforced brick masonry; and (5) verifying the suitability of a strength method for brick masonry beams with various percentages of steel. The study is limited to rectangular flexural members of brick masonry using Type S portland cement lime mortar. A low compressive strength and a high compressive strength brick unit were examined for loading both parallel and perpendicular to the bed joints.

Test Program

The Construction Research Center at the University of Texas at Arlington has conducted a two phase experimental research investigation to develop, if feasible, an ultimate flexural strength design procedure for reinforced brick masonry. Specimens were composed of extruded brick units (low and high strength) and Type S (1C, ½L, 4¼S) mortar. In normal masonry construction, beam elements produce compression parallel to the bed joints whereas wall elements typically exhibit compression perpendicular to bed joints. Both compression conditions for both brick types were examined and verified by test data.

Phase I consisted of testing 20 beam column specimens and corresponding prisms to determine the actual flexural compressive stress distribution for (*a*) high and low strength brick masonry and for (*b*) compression parallel and perpendicular to bed joints. The results have established the parametric values needed for defining a stress block for brick masonry at ultimate strength conditions.

Phase II consists of reinforced brick masonry flexural tests (still in progress) to substantiate that the ultimate strength concept is applicable to reinforced flexural elements with varying percentages of steel. Two strength brick units with parallel to bed joint compression stress and three different steel percentages are being examined.

Materials

Brick

Two types of bricks as classified by their average compressive strength were used in making specimens. The cored three-holed, low strength unit measured 5.72 by 8.89 by 20.64 cm (2¼ by 3½ by 8⅛ in.) with an initial rate of absorbtion (IRA) of 53.2 g/min and a compressive strength of 27 280 kPa (3957 psi) as tested according to ASTM Standard for Sampling and Testing Brick and Structural Clay Tile (C 67). The solid high strength brick measuring 5.40 cm by 8.57 cm 19.05 cm (2⅛ in. by 3⅜ in. by 7½ in.) exhibited an IRA of 0.6 g/min and a compressive strength of 106 905 kPa (15 507 psi).

Grout

Grout was used in accordance with ASTM Specification for Grout for Reinforced and Unreinforced Masonry (C 476-80), with a slump of 25.4 cm (10 in.).

Steel

Grade 60 deformed steel bars were used for all reinforced elements. The average proportional limit, yield plateau, and ultimate tensile strength were 396 405, 482 580, and 486 027 kPa, (57.5, 70, and 70.5 ksi), respectively.

Other Materials

Portland cement Type I was certified to be in compliance with ASTM Specification for Portland Cement (C 150-81). Lime was certified to be in compliance with ASTM Specification for Hydrated Lime for Masonry Purposes (C 207-79). Masonry sand, concrete sand and pea gravel conformed to the sieve analysis requirements in ASTM Specification for Aggregate for Masonry Mortar (C 144-81) and ASTM Specification for Aggregates for Masonry Grout (C404-81).

Mixing

All ingredients were proportioned by volume. Mortar mixing was done in a 0.14 m^3 (5 ft^3) paddle type mixer for a period of 5 min to the desired consistency of the mason. Grout mixing was done in a 0.17 m^3 (6 ft^3) capacity rotary concrete mixer.

Phase I—Ultimate Compressive Stress Distributions

Development of Test Method

Consider an unreinforced masonry member as shown in Fig. 1*a*. The couple moment M and axial force P can be arranged so that the stress distribution in the masonry section is as shown in Fig. 1*b*; i.e., the entire cross section is in compression with zero stress on one face.

By equilibrium of forces

$$P = k_1 k_3 f'm\, bc$$

$$k_1 k_3 = \frac{P}{f'mbc}$$

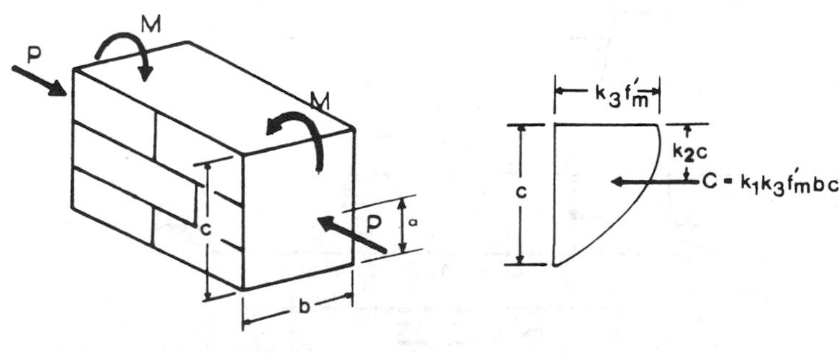

FIG. 1—*Proposed experimental stress distribution.*

By equilibrium of moments

$$Pa + M = (1 - k_2)k_1k_3 f'm\, bc^2$$

$$k_2 = 1 - \frac{Pa + M}{Pc}$$

The important parameters k_1k_3 and k_2 can be computed directly from a measured force P and an applied moment M from zero load up to failure without any assumptions on the actual stress distribution in the masonry.

In 1954, Hognestad [*1*] successfully conducted an investigation to develop the flexural compressive concrete stress block using specimens as shown in Fig. 2. Applying a major thrust P_1 and a minor thrust P_2 the neutral axis can be maintained at one face of the test specimen. For this loading system:

$$k_1k_3 = \frac{P_1 + P_2}{f'mbc} = \frac{f_0}{f'm}$$

$$k_2 = 1 - \frac{P_1a_1 + P_2a_2}{(P_1 + P_2)c} = 1 - \frac{m_0}{f_0}$$

where

$$f_0 = \frac{P_1 + P_2}{bc}$$

and

$$m_0 = \frac{P_1a_1 + P_2a_2}{bc^2}$$

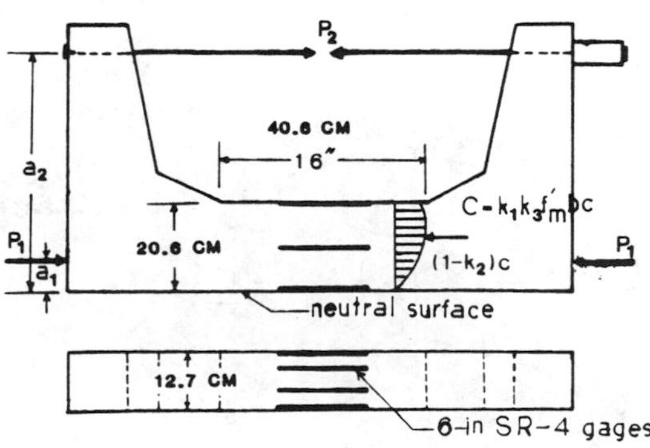

FIG. 2—*Concrete test specimen.*

The compressive flexural stress-strain relationship for this loading condition can be determined by Fig. 3 from the following assumptions:

1. A plane section before loading remains a plane section after loading. This results in a linear strain variation across the section, which can be verified by strain measurements.
2. Masonry stress is a function of strain only.

From Figs. 2 and 3 by equilibrium of forces

$$C = b\int_0^c F(\epsilon_x)\,dx = P_1 + P_2$$

By equilibrium of moments

$$M = b\int_0^c F(\epsilon_x)x\,dx = P_1a_1 + P_2a_2$$

Using the above assumptions, these force and moment relations lead to

$$f_c = f_0 + \epsilon_c\, df_0/d\epsilon_c \quad (1)$$

and

$$f_c = 2m_0 + \epsilon_c \frac{dm_0}{d\epsilon_c} \quad (2)$$

Equations 1 and 2 give the extreme flexural compressive stress f_c as a function of the measured extreme flexural compressive strain ϵ_c and the measured valves f_0 and m_0. The values of $df_0/d\epsilon_c$ and $dm_0/d\epsilon_c$ can be determined from plots of the measured loads and ϵ_c. By comparing the relationship between f_c and ϵ_c by Eq 1 and 2 a check on the accuracy of the test data is made. From this information a complete flexural stress-strain curve at an ultimate load can be generated.

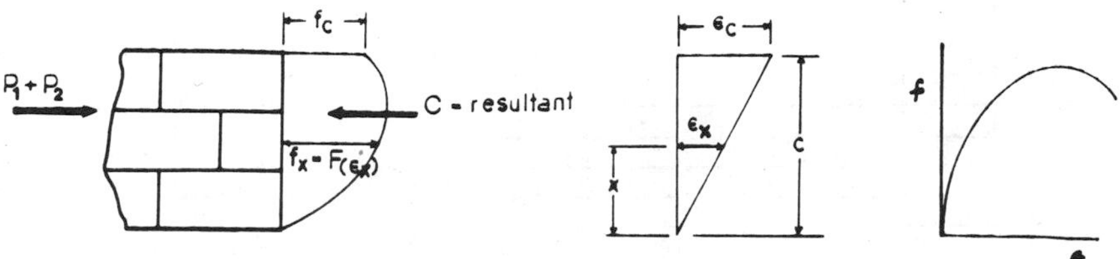

FIG. 3—*Stress and strain distributions.*

The first preliminary masonry test specimens were built as shown in Fig. 4. The reinforced masonry end brackets, built for transfering the minor thrust P_2, performed unsatisfactorily. The second test specimens were built without masonry brackets as shown in Fig. 5*a*. Steel brackets as shown in Fig. 5*b* were designed to transmit thrust P_2 to the test specimen.

The specimens for loading parallel to the bed joint (25.4 mm or 10 in. wide) were built of two brick wythes with a center grout core. Each wythe had a three-course running bond pattern, seven brick long with 9.5 mm (3/8 in.) tooled mortar joints. Companion prisms were built 2½ brick long with an identical cross section. See Fig. 6*a*.

The specimens for loading perpendicular to the bed joints (25.4 mm or 10 in. wide) were built using a two-wythe soldier course, 21 brick long with 9.5 mm (3/8 in.) tooled mortar joints and a center grout core. See Fig. 6*b*.

All specimens were air cured in the laboratory a minimum of 28 days after grouting before testing.

Test Equipment

The beam column specimens were tested to ultimate using the loading arrangement shown in Fig. 7. The major thrust P_1 was applied by a 181 t (400 000 lb) hydraulic testing machine. The minor thrust P_2 was applied by a 27 t (60 000 lb) hydraulic ram. Compression strains were measured at the neutral surface, the mid-depth, and the extreme compression surface by placing three linear variable-displacement transducers over a 20.3 cm (8 in.) gage length on two opposite sides of the test specimen. The six transducer measurements were recorded using an automatic scan strain indicator system and averaged. The companion prism specimens were tested to ultimate according to ASTM Test Methods for Compressive Strength of Masonry Prisms (E 447), recording compressive strain

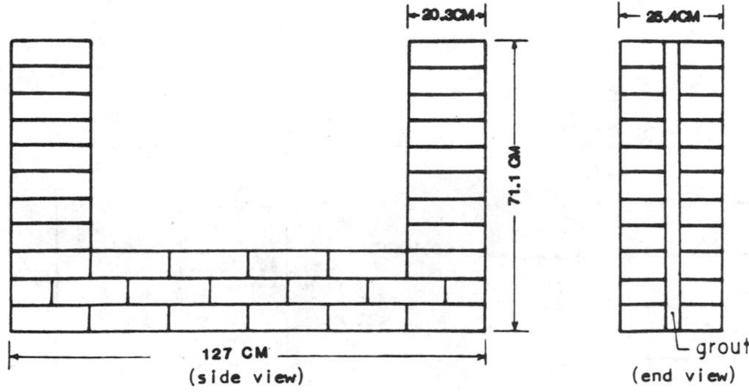

FIG. 4—*Phase 1 preliminary test specimen.*

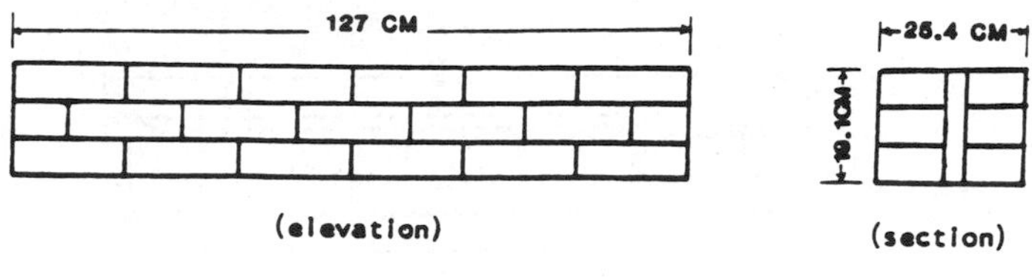

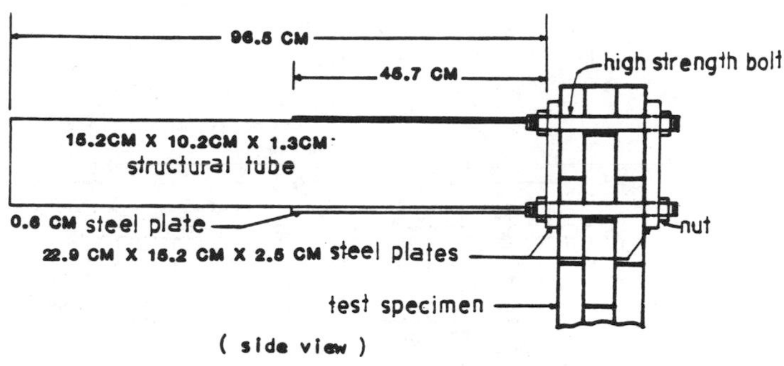

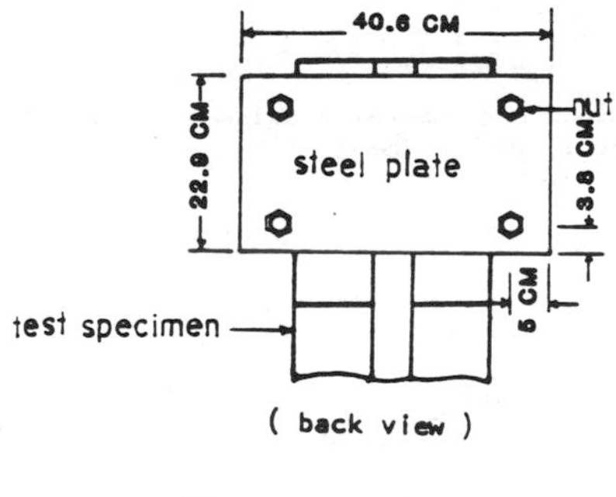

FIG. 5(a)—*Phase 1 final masonry test specimen.* (b) *Phase 1 loading bracket.*

measurements using transducers over a 20.3 cm (8 in.) gage distance at each of the prism's four corners.

Test Procedure

The major thrust P_1 was applied to desired increment. The minor thrust P_2 was then applied by jacking against the reaction frame to achieve zero readings

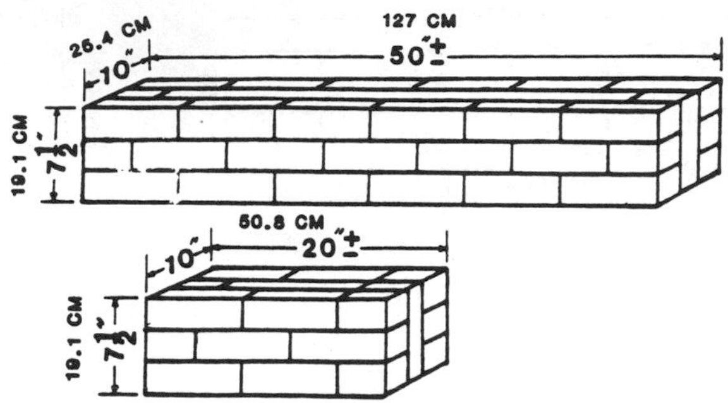

a) Test specimen and companion prism for loading parallel to bed joint.

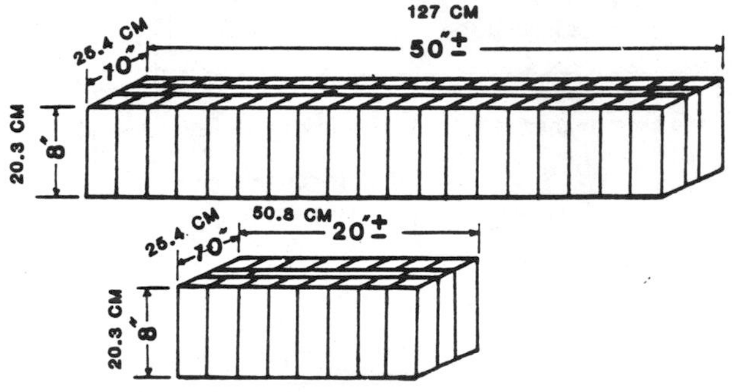

b) Test specimen and companion prism for loading perpendicular to bed joint.

FIG. 6—*Phase 1 test specimens and prisms.*

in the transducers at the neutral surface, at which time compressive strain readings were recorded. This sequence was continued to failure. The major thrust P_1 increments for the low strength masonry began at 4545 kg (10 kips). At approximately three-fourths of the maximum load the increment was reduced to 2273 kg (5 kips). For the high strength brick specimens, the initial P_1 increment was 9090 kg (20 kips). At half the maximum load the increment was reduced to 4545 kg (10 kips). At three-fourths of the maximum load the increment was changed to 2273 kg (5 kips).

Beam Column Behavior

For compression both parallel and perpendicular to the bed joints the low strength brick assemblages showed similar behavior. No sign of cracking was noticed until the maximum load was reached. Failure started gradually with

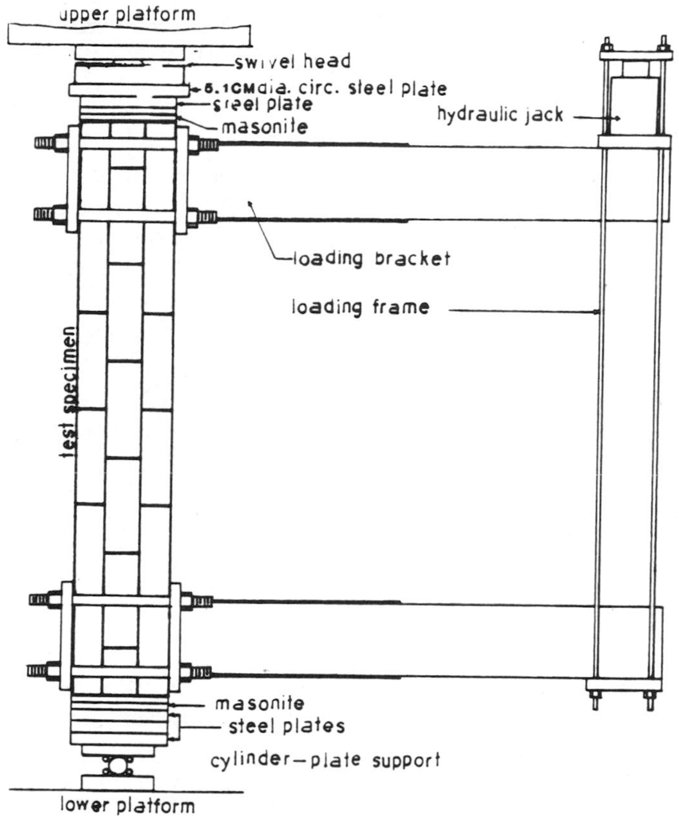

FIG. 7—*Phase 1 load test arrangement.*

crushing of material on the face of maximum compressive strain. Crushing slowly progressed toward the neutral surface while both the major and minor thrusts were decreasing.

For loading perpendicular to the bed joints in the high strength brick specimens, hairline cracks occurred along the contact surface between the brick and grout on the face in maximum compression at a load between one-third and one-half of the ultimate. These cracks continued to widen up to a load of three-fourths of the maximum. No cracks were observed in the individual masonry wythe joints before crushing occurred. Two of the specimens exhibited a sudden explosive failure. The remaining specimens showed crushing with a sudden drop in load.

For loading parallel to the bed joints, the high strength brick specimens showed continuous cracks in every bed joint and between the brick and grout at one-third of the ultimate capacity. These cracks widened upon increase in load. Close to ultimate, cracks in the order of 3.2 mm (1/8 in.) were observed between the grout and the brick and in the bed joints on the maximum compression face. Crushing of the brick and grout was followed by a drop in loading.

Test Results and Analysis

Figures 8 through 11 show the average strain profile of beam column specimens across the masonry cross section at different loads. These curves confirm the accuracy of the assumption that plane sections remain plane.

The average maximum compressive masonry strain for low strength brick specimens loaded perpendicular to the bed joints was 2452×10^{-6} under an average major thrust P_1 of 60.7 t (133.75 kips). Data indicates a 5.6% coefficient of variation and 3.4 t (7.5 kips) standard deviation for P_1.

The average maximum masonry compressive strain for low strength brick specimens loaded parallel to the bed joints was 2372×10^{-6} under an average major thrust P_1 of 44.5 t (98 kips). Data shows a 5.8% coefficient of variation and a 2.6 t (5.7 kips) standard deviation for P_1.

The average maximum masonry compressive strain for high strength brick specimens loaded perpendicular to the bed joint was 5470×10^{-6} under an average major thrust of P_1 of 124.5 t (274 kips). Data indicates a 3.5% coefficient of variation and 4.4 t (9.6 kips) standard deviation for P_1.

The average maximum masonry compressive strain for high strength brick specimens loaded parallel to the bed joints was 3397×10^{-6} under an average major thrust P_1 of 122.7 t (270 kips). Data gives a 2.3% coefficient of variation with 2.8 t (6.1 kips) standard deviation for P_1.

Both strength beam column brick specimens showed higher strength and more ductility for loading perpendicular to bed joints.

Figures 12 through 15 show the stress-strain relationships obtained from axial prism tests. For both brick strengths, higher average assemblage strengths are found for loading perpendicular to bed joints.

Figures 16 through 19 give the measured values of f_0 and m_0 as a function of the measured strain on the masonry specimens' maximum compression face. This in conjunction with Eqs 1 and 2 generates the flexural compressive stress-strain curves shown in Fig. 20. The closeness of these curves indicates the accuracy of the test method. Averaging f_c from both equations results in the average flexural compressive stress-strain curves in Figs. 21 through 24. These figures also give a comparison between the axial and flexural values.

Figures 25 through 28 give measured values of k_2 and k_1k_3 as functions of maximum flexural compressive strain from zero loading to failure for each group of beam column specimens. The k_1k_3 curves are smooth and increase monotonically with strain. The k_2 curves fluctuate at the beginning of loading, then quickly merge together to a nearly constant value at failure. Individual values of k_2 and k_1k_3 at ultimate load are given in Tables 1 and 2.

Table 1 gives a comparison for the stress block parameters. In this table, $A \times b$ is the area under the generated stress-strain curve times the width of the beam

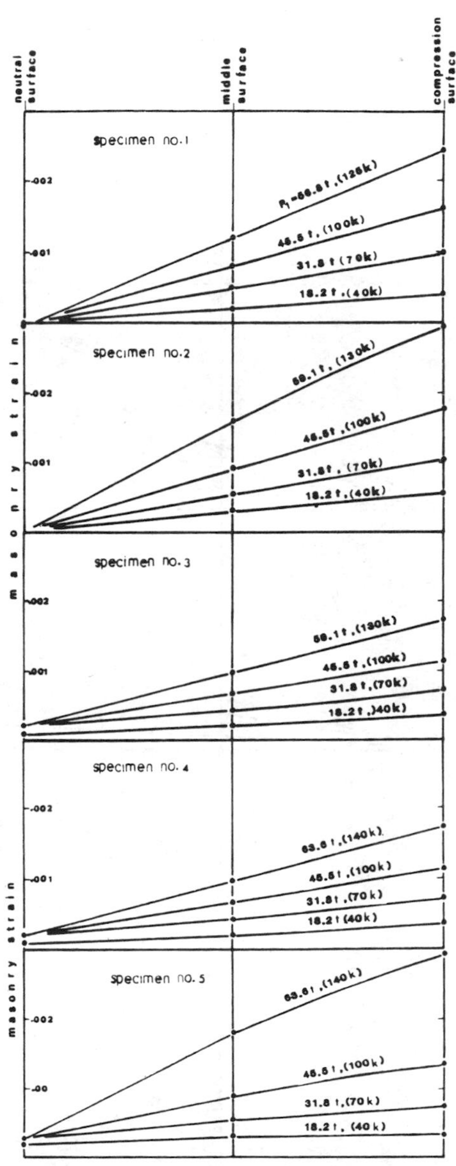

FIG. 8—*Strain profiles at various applied loads for low-strength brick masonry specimens loading normal to bed joint.*

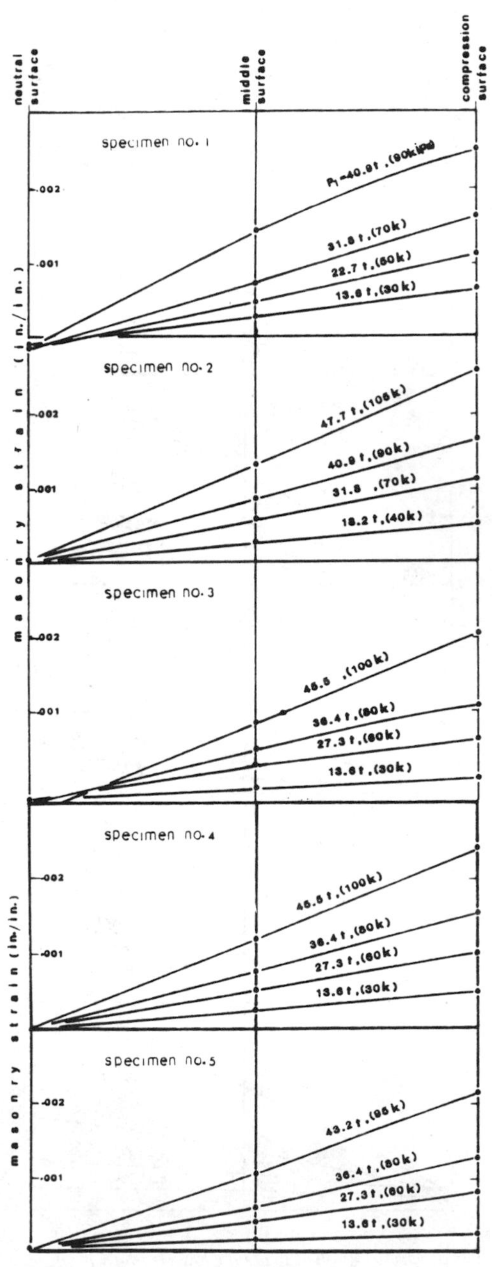

FIG. 9—*Strain profiles at various applied loads for low-strength brick masonry specimens loading parallel to bed joint.*

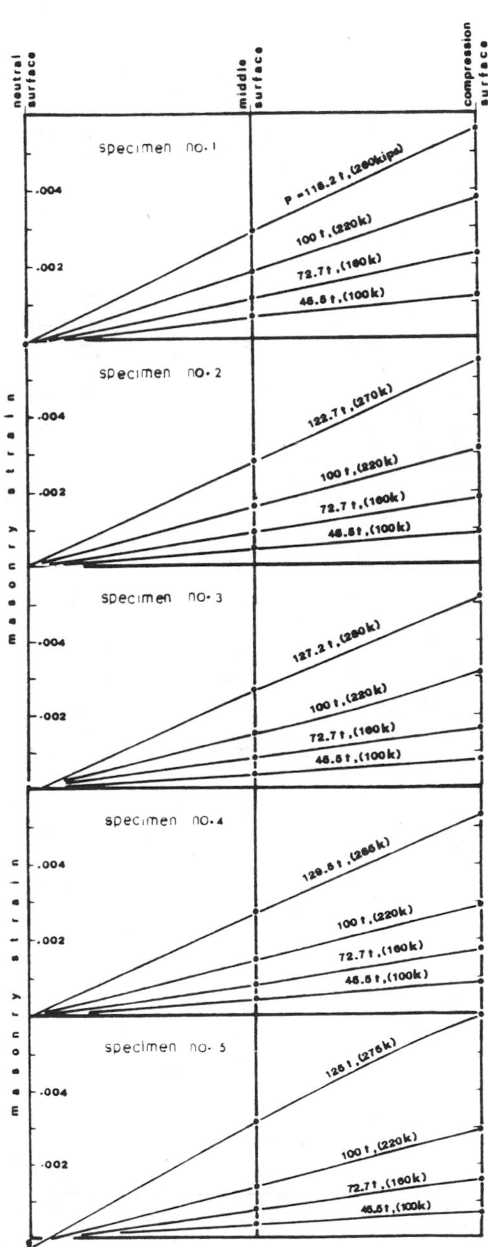

FIG. 10—*Strain profiles at various applied loads for high-strength brick masonry specimens loaded normal to bed joint.*

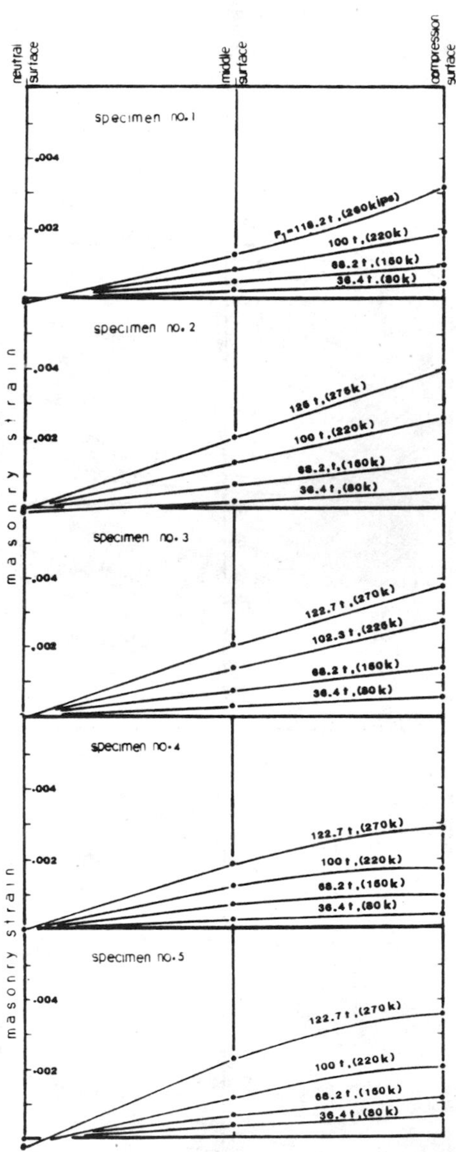

FIG. 11—*Strain profiles at various applied loads for high-strength brick masonry specimens loaded parallel to bed joint.*

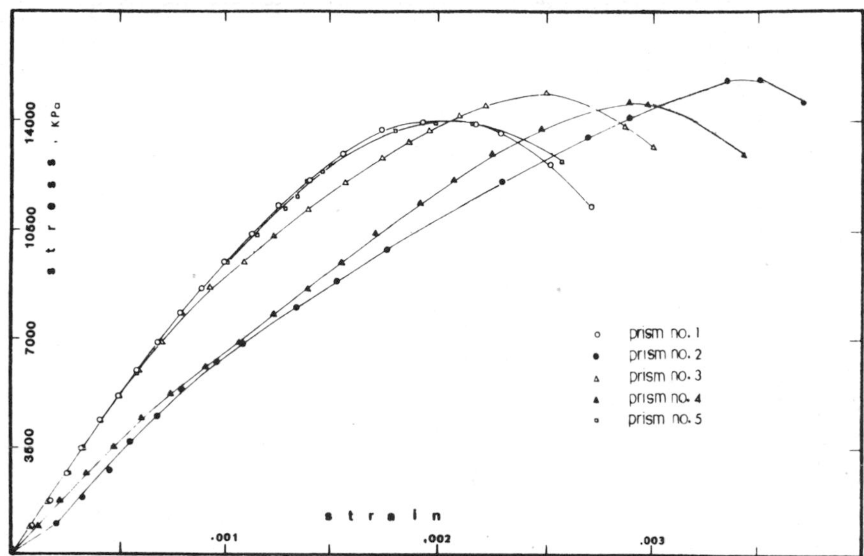

FIG. 12—*Axial stress-strain relationship for low-strength brick masonry prisms loading perpendicular to bed joint.*

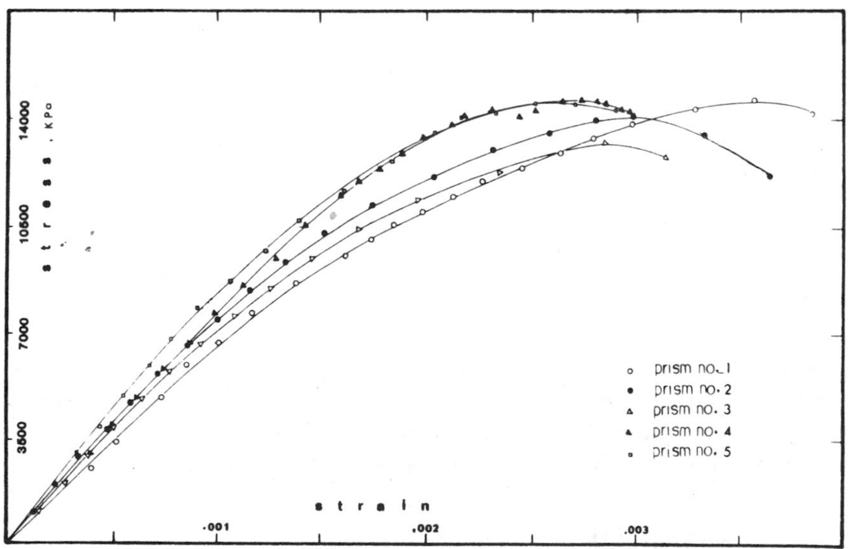

FIG. 13—*Axial stress-strain relationship for low-strength brick masonry prisms loading parallel to bed joint.*

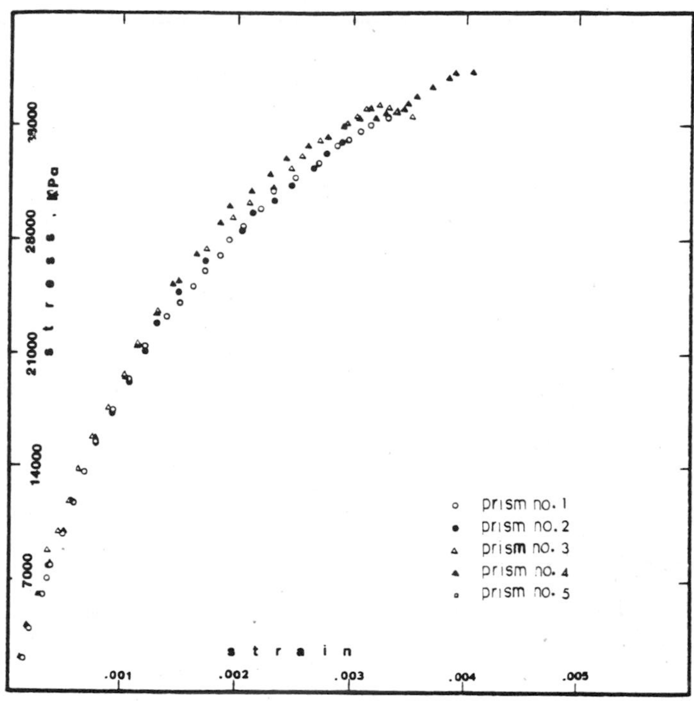

FIG. 14—*Axial stress-strain relationship for high-strength brick masonry prisms loading perpendicular to bed joint.*

column specimen. Good agreement exists between the total compressive force applied to each beam column specimen ($P_1 + P_2$) and the internal resultant compressive force. This validates Eqs 1 and 2 and indicates the accuracy of $k_1 k_3$. Values of k_2 can be determined from the external applied load condition (measured) and from the centroid of the area under the developed stress-strain curve. A comparison between the two methods of computing k_2 shows disagreement in the order of 3 to 8%. The average values from the two methods are given in Table 2.

The ultimate compressive flexural strains from test results of 20 beam column specimens and 9 reinforced beams (to be reported) are plotted against companion axial prism strength tests in Fig. 29. The graph shows no distinction between the two different directions of compressive stress to bed joints. The relationship is represented by two discontinuous straight line equations.

Phase 2—Reinforced Brick Masonry Beams

To evaluate the accuracy of the parameters developed in Phase 1 and a proposed ultimate strength design procedure, reinforced masonry beams were

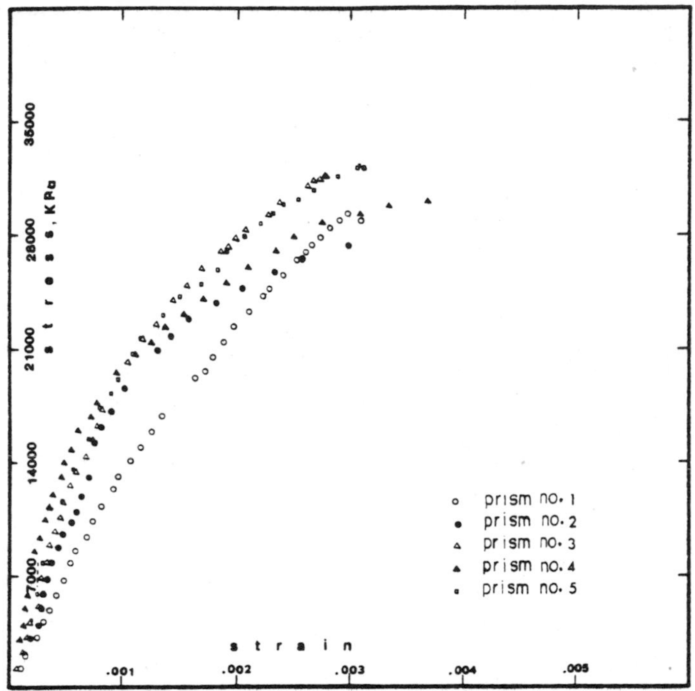

FIG. 15—*Axial stress-strain relationship for high-strength brick masonry prisms loading parallel to bed joint.*

constructed using Phase 1 low strength brick. Grade 60 reinforcement of various percentages were approximated to generate compression failure, balanced failure, and tension failure for compression parallel to bed joints. Three companion beams per cross section as shown in Fig. 30 along with companion prisms were constructed. Transducers measured longitudinal strains at midspan ½ in. from beam bottom, at center of beam, and at top of beam. Vertical deflections were measured at midspan. The beam loading brackets are shown in Fig. 31; the test setup in Fig. 32. These test results, along with additional high strength brick beam tests, are currently being evaluated along with existing published beam strength tests in the generation of a proposed flexural strength design method for reinforced brick masonry. This information will be published in the spring of 1984.

At this time the following tentative findings are drawn:

1. An ultimate strength theory reasonably predicts the flexural capacity of brick masonry.

2. An equivalent rectangular stress block can be used in lieu of the actual flexural stress distribution.

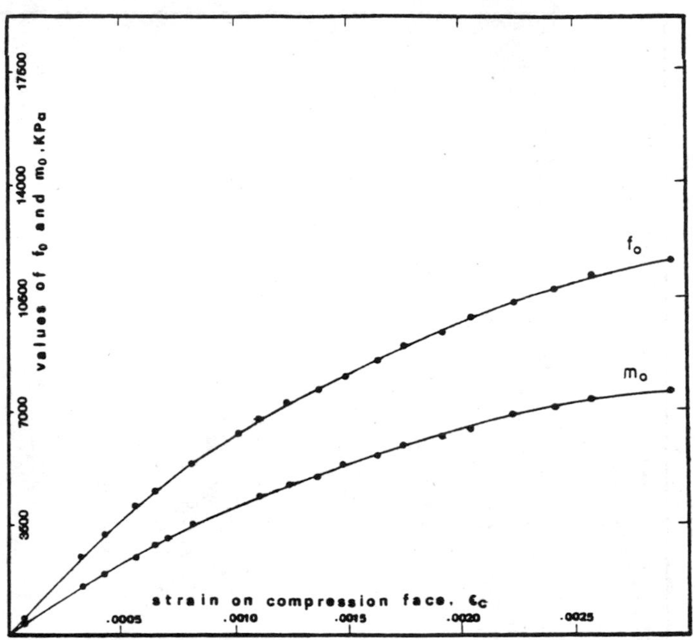

FIG. 16—f_0 *and* m_0 *as functions of strain on compression face for specimen No. 2 low-strength load perpendicular to bed joint.*

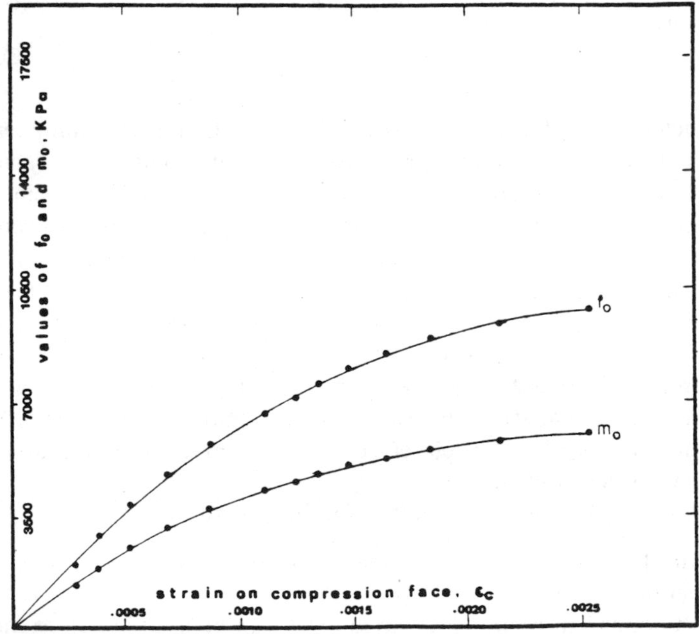

FIG. 17—f_0 *and* m_0 *as functions of strain on compression face for specimen No. 2 low-strength load parallel to bed joint.*

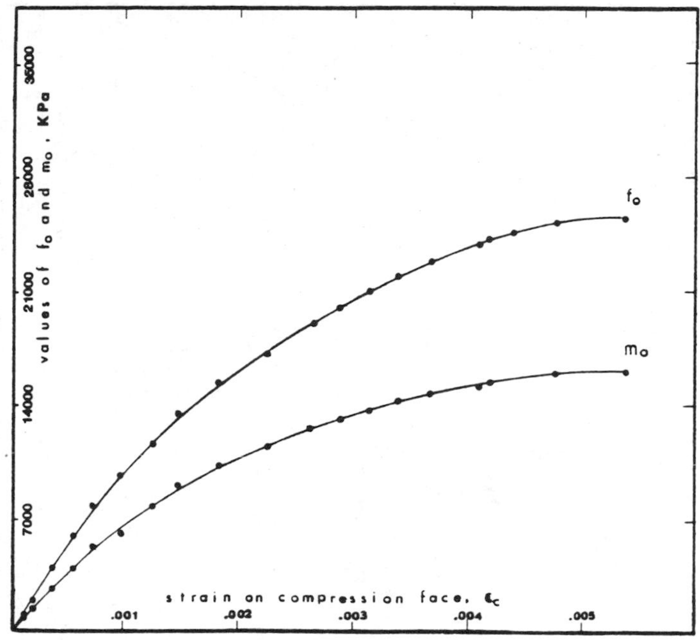

FIG. 18—f_0 *and* m_0 *as functions of strain on compression face for specimen No. 2 high-strength load perpendicular to bed joint.*

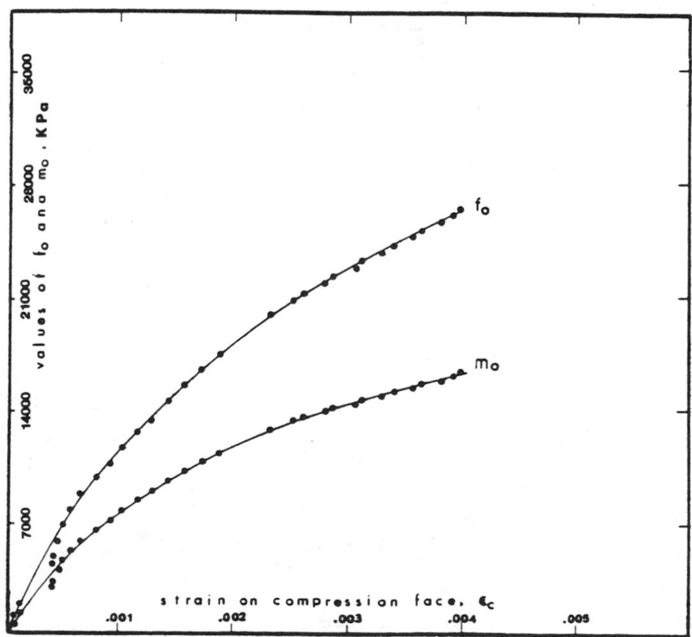

FIG. 19—f_0 *and* m_0 *as functions of strain on compression face for specimen No. 2 high-strength load parallel to bed joint.*

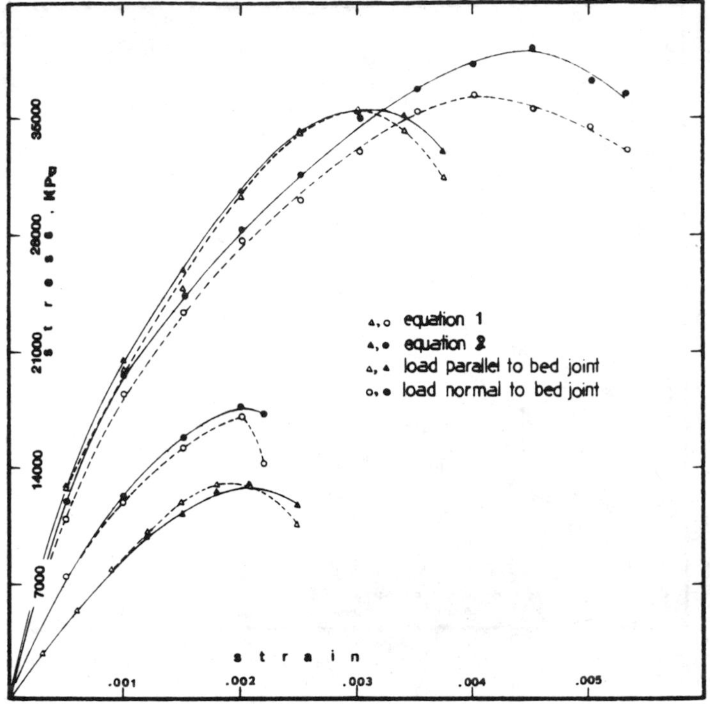

FIG. 20—*Flexural compressive stress-strain curves.*

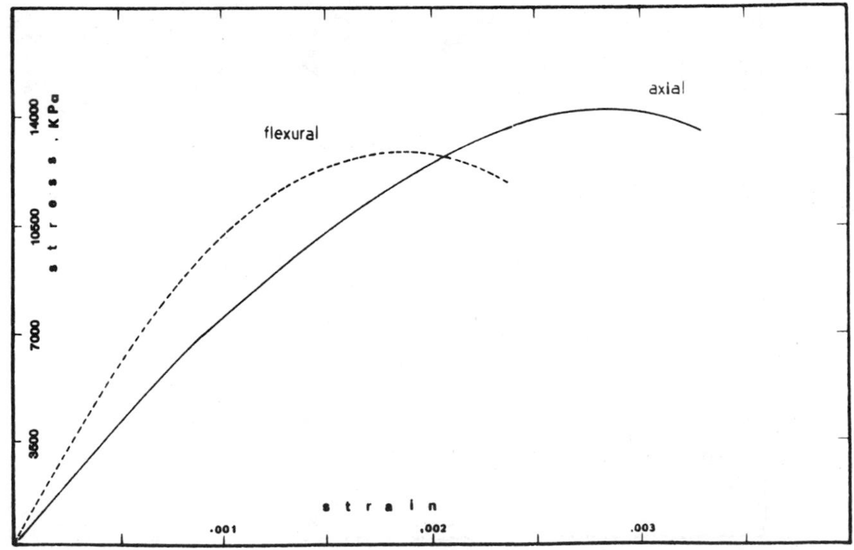

FIG. 21—*Comparison between average flexural and axial stress-strain relationship for low-strength brick masonry with compression parallel to bed joint.*

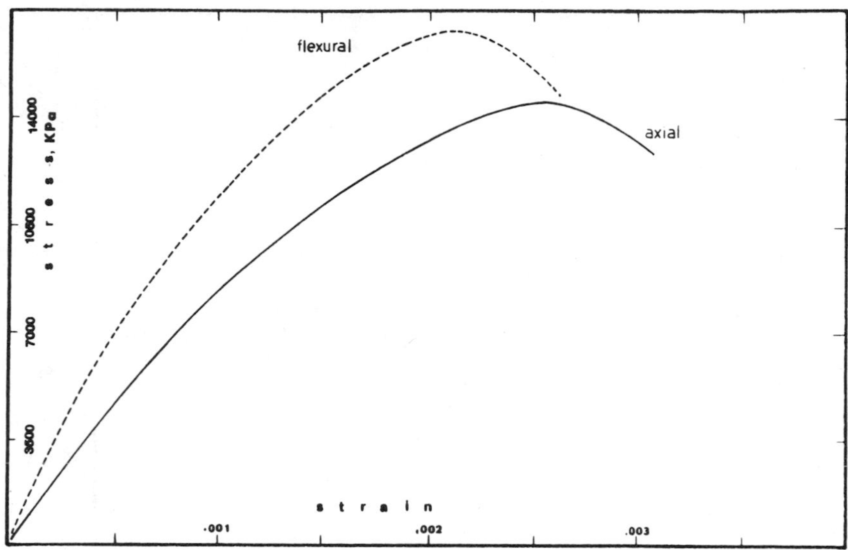

FIG. 22—*Comparison between average flexural and axial stress-strain relationship for low-strength brick masonry with compression perpendicular to bed joint.*

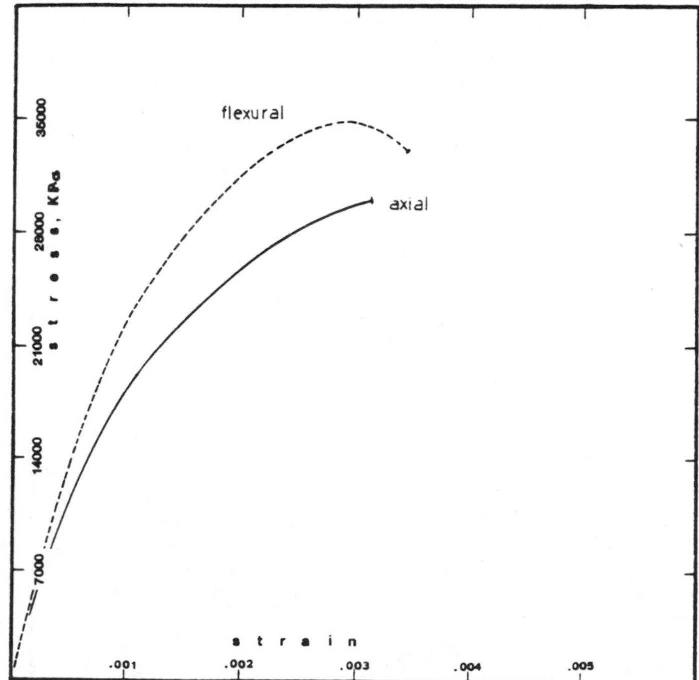

FIG. 23—*Comparison between average flexural and axial stress-strain relationship for high-strength brick masonry with compression parallel to bed joint.*

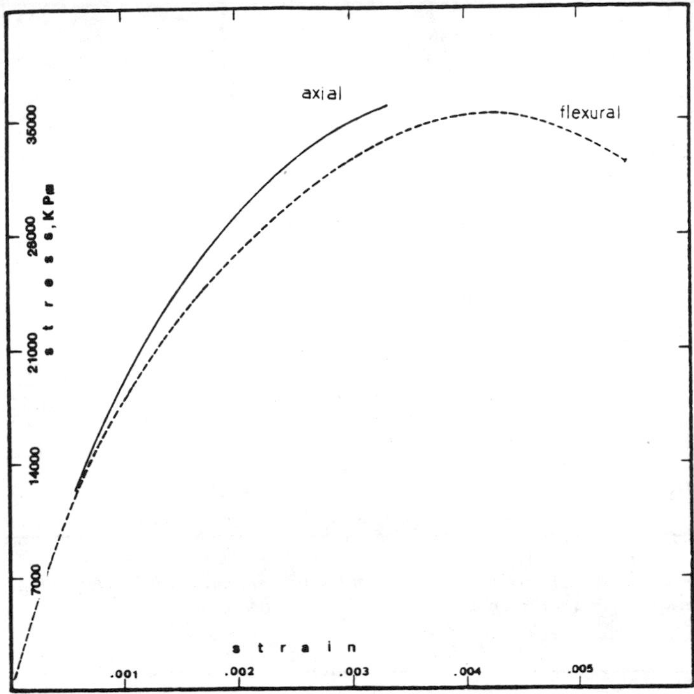

FIG. 24—*Comparison between average flexural and axial stress-strain relationship for high-strength brick masonry with compression perpendicular to bed joint.*

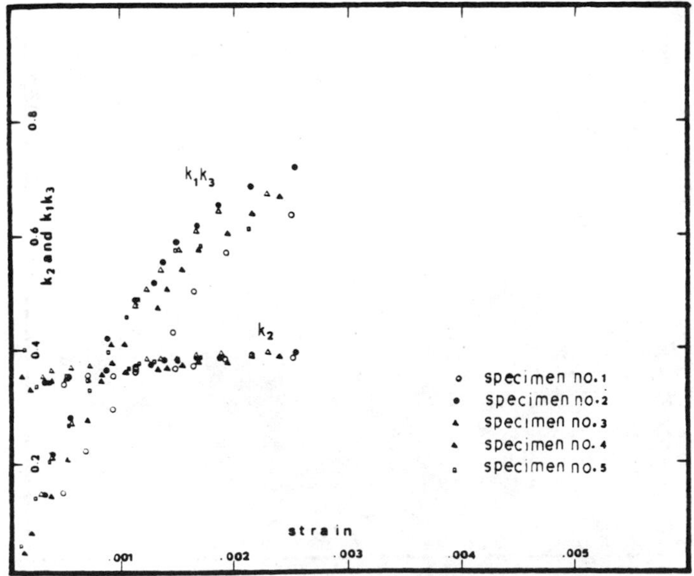

FIG. 25—*Values of* k_2 *and* k_1k_3 *at various stress levels as functions of strain on compression face for low-strength brick masonry loading parallel to bed joint.*

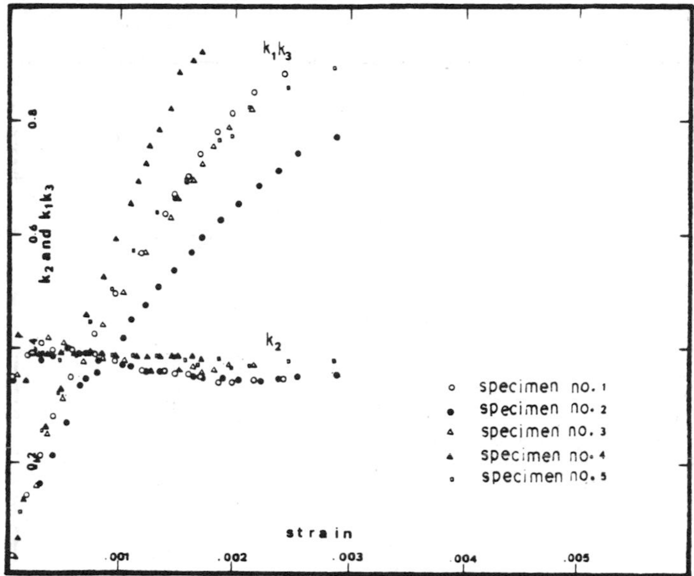

FIG. 26—*Values of* k_2 *and* k_1k_3 *at various stress levels as functions of strain on compression face for low-strength brick masonry loading perpendicular to bed joint.*

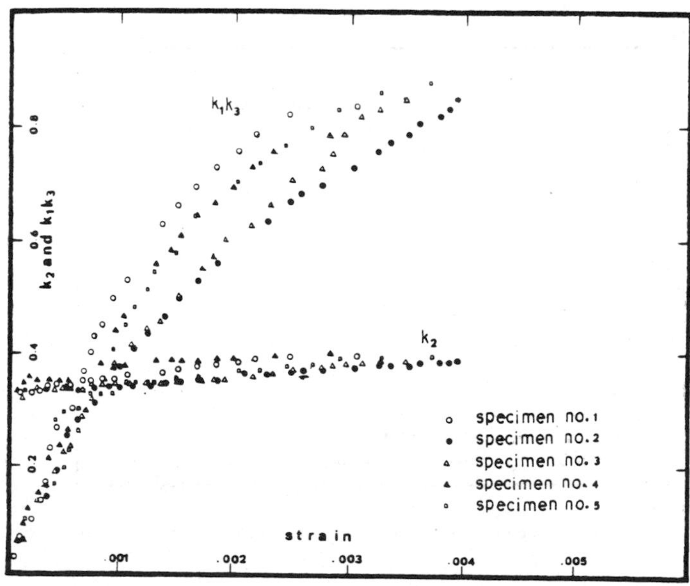

FIG. 27—*Values of* k_2 *and* k_1k_3 *at various stress levels as functions of strain on compression face for high-strength brick masonry loading parallel to bed joint.*

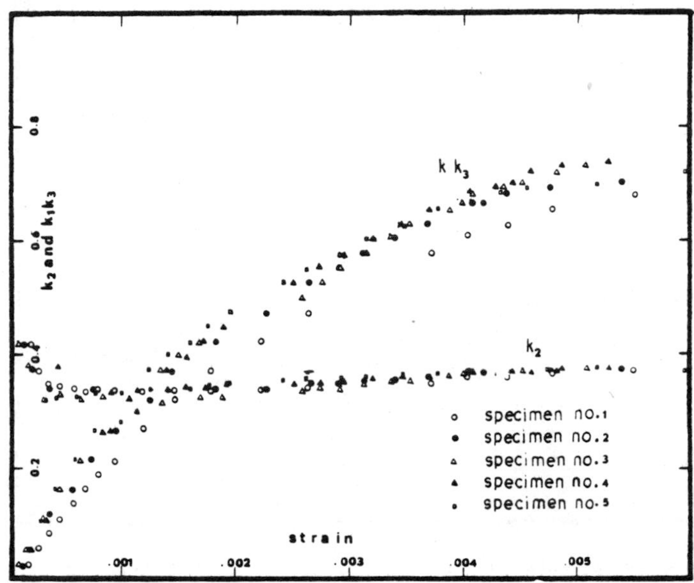

FIG. 28—*Values of* k_2 *and* k_1k_3 *at various stress levels as functions of strain on compression face for high-strength brick masonry loading perpendicular to bed joint.*

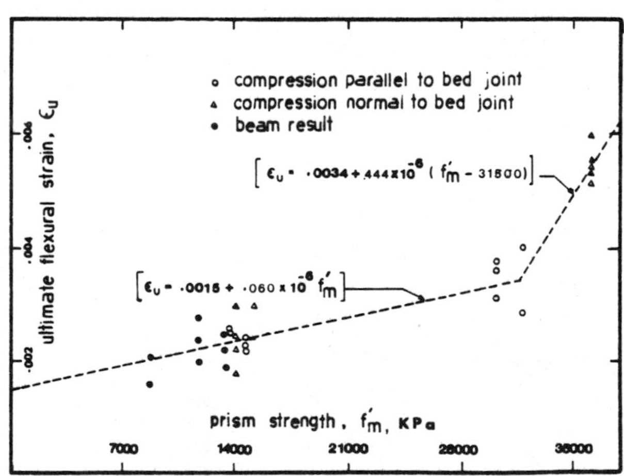

FIG. 29—*Relationship between ultimate flexural strain and the corresponding prism strength.*

TABLE 1—*Properties of flexural stress block determined from curves as compared to measured values.* A × b *is the area under the curve times width of the column.*

Specimen Group	Column	$\frac{A \times b}{P_1 + P_2}$	Values of k_2	
			Curve	Measured
Low-strength load parallel to bed joint	1	1.002	0.379	0.380
	2	1.026	0.417	0.390
	3	0.924	0.401	0.391
	4	1.008	0.395	0.381
	5	1.013	0.421	0.382
	avg	0.9946	0.4026	0.3848
Low-strength load perpendicular to bed joint	1	1.025	0.388	0.346
	2	1.040	0.396	0.354
	3	1.005	0.386	0.370
	4	0.988	0.373	0.375
	5	1.033	0.415	0.377
	avg	1.0182	0.3916	0.3644
High-strength load parallel to bed joint	1	1.026	0.431	0.396
	2	1.008	0.409	0.385
	3	1.022	0.410	0.385
	4	1.021	0.421	0.397
	5	1.009	0.414	0.390
	avg	1.017	0.417	0.3906
High-strength load perpendicular to bed joint	1	1.026	0.402	0.370
	2	1.043	0.397	0.374
	3	1.008	0.409	0.374
	4	1.025	0.407	0.370
	5	1.040	0.426	0.373
	avg	1.0284	0.4082	0.3722

3. Deflections are reasonably predicted using the current ACI code procedure at working loads for beams constructed with low strength units and with balanced or under reinforced percentages of steel.

Summary and Conclusions

Twenty beam column specimens along with companion prisms were constructed of high and low strength brick masonry units for compression loading both parallel and perpendicular to bed joints for generation of the parameters defining the ultimate flexural stress distribution. Underreinforced, balanced, and overreinforced brick masonry beams of two brick grades were constructed to evaluate the accuracy of the stress block parameter. These tests are still in progress. This data along with existing data are being evaluated in light of an

TABLE 2—*Values of* $k_1 k_3$ *and individual values of* k_1, k_2 *and* k_3 *at ultimate loads.*

Specimen Group	Column	f_{max}, psi	$f'm$, psi	$k_1 k_3$	k_3	k_1	k_2
Low-strength load parallel to bed joint	1	1 824	1 964	0.637	0.929	0.686	...
	2	1 851	1 964	0.719	0.942	0.763	...
	3	1 914	2 097	0.674	0.913	0.738	...
	4	1 965	2 097	0.670	0.937	0.715	...
	5	1 708	2 097	0.610	0.814	0.749	...
	avg	1 852	2 044	0.662	0.907	0.730	0.3937
Low-strength load perpendicular to bed joint	1	2 652	2 009	0.891	1.32	0.675	...
	2	2 449	2 156	0.776	1.136	0.683	
	3	2 472	2 009	0.824	1.230	0.670	...
	4	2 975	2 009	0.976	1.481	0.625	...
	5	2 481	2 009	0.897	1.235	0.726	...
	avg	2 606	2 038	0.863	1.280	0.676	0.3780
High-strength load parallel to bed joint	1	4 645	4 297	0.857	1.081	0.793	
	2	5 376	4 526	0.849	1.188	0.715	...
	3	5 045	4 297	0.846	1.174	0.721	...
	4	4 737	4 526	0.788	1.047	0.753	...
	5	4 988	4 297	0.878	1.163	0.755	...
	avg	4 960	4 389	0.844	1.131	0.747	0.4038
High-strength load perpendicular to bed joint	1	4 925	5 135	0.692	0.959	0.722	...
	2	4 986	5 135	0.718	0.971	0.739	...
	3	5 308	5 135	0.745	1.034	0.721	...
	4	5 336	5 135	0.759	1.039	0.731	...
	5	4 819	5 135	0.731	0.938	0.779	...
	avg	5 075	5 135	0.729	0.988	0.738	0.3902

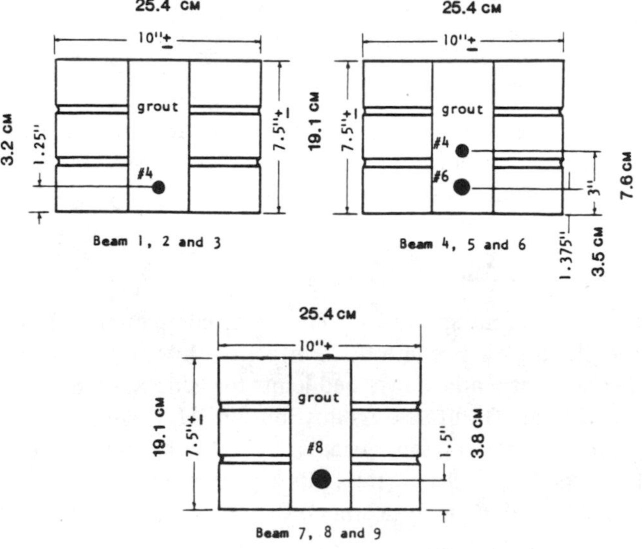

FIG. 30—*Cross-sectional details for low-strength brick masonry beams.*

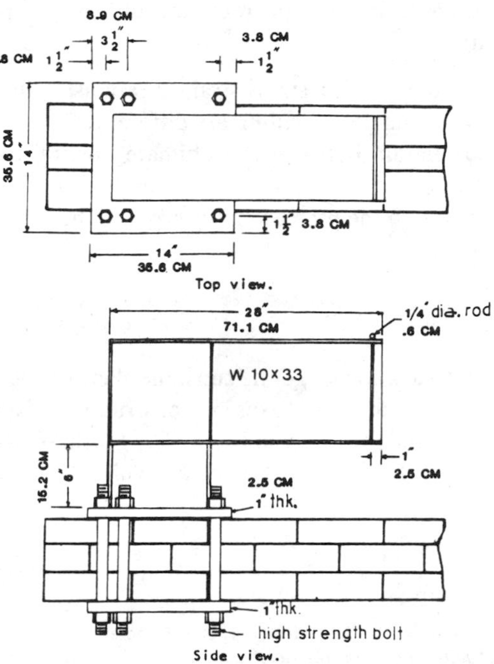

FIG. 31—*Loading bracket equipped to test beam.*

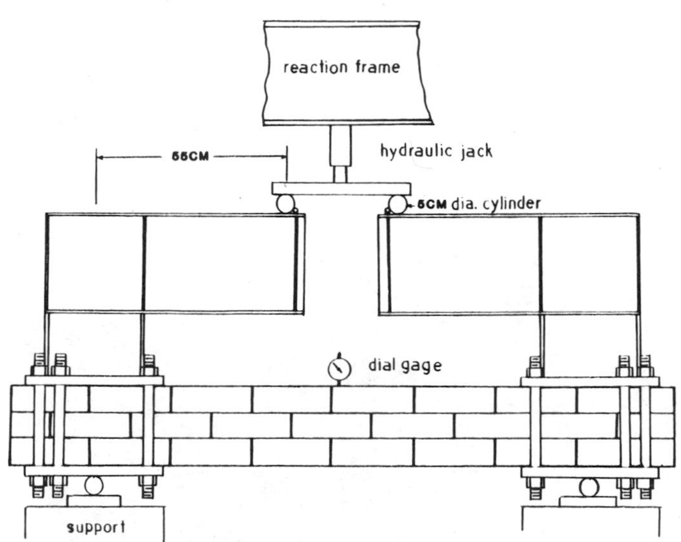

FIG. 32—*Phase II test setting up.*

ultimate strength theory. Based on current data and analysis the following conclusions are drawn:

1. The ultimate compressive stress-strain relationship for brick masonry under flexural loading and axial loading are different.
2. There is a correlation between the ultimate flexural compressive strain and prism strength.
3. An ultimate strength theory reasonably predicts the capacity of reinforced brick masonry beams.

Acknowledgments

The authors wish to acknowledge the substantial efforts of Mr. Don Halsell, director of The Brick Institute of Texas, in construction of all test specimens. All brick units were donated by Acme Brick Co., Forth Worth, TX. Financial support from the Brick Institute of America permitted the project concept to become a reality.

References

[*1*] Hognestad, E., Hansen, N. W., and McHenry, D., "Concrete Stress Distribution in Ultimate Strength Design," *Journal of The American Concrete Institute*, Dec. 1955.

Clayford T. Grimm[1]

Corrosion of Steel in Brick Masonry

REFERENCE: Grimm, C. T., **"Corrosion of Steel in Brick Masonry,"** *Masonry: Research, Application, and Problems, ASTM STP 871*, J. C. Grogan and J. T. Conway, Eds., American Society for Testing and Materials, Philadelphia, 1985, pp. 67–87.

ABSTRACT: Under present design standards steel ties, anchors, fasteners, joint reinforcement, reinforcing bars, and structural steel lintels and shelf angles in exterior masonry walls corrode at a rate which gives them a life expectancy of 5 to 70 years. Implementation of recommendations for improved design standards could double their life expectancy. Seventy references are cited.

KEY WORDS: anchor (masonry), beam, brick, building, column, corrosion, galvanizing, lintel, masonry, reinforcing (bar), shelf angle, steel, tie (masonry), wall

The structural performance of brick masonry is usually dependent on the strength of metal connectors, reinforcement, and/or structural supports. Connectors include ties, anchors, and fasteners [*29*].[2] Reinforcing bars are often placed in masonry to provide increased strength. Structural supports include lintels and shelf angles.

Ties hold parts (that is, wythes) of masonry walls together. Anchors attach masonry walls to their lateral supports, for example, structural frames. Fasteners attach building fixtures to walls, for example, fasteners for chalkboards. The dead load of masonry above wall openings is usually carried by lintels or shelf angles. Shelf angles may also support the dead load of nonbearing masonry panel walls on structural frames. Lintels may be of reinforced masonry [*51*] or structural steel shapes [*63*]. Shelf angles are usually of structural steel [*4*]. Steel is the most common material for connectors, reinforcement, and structural supports.

Corrosion of steel occurs in the presence of water and oxygen and is accelerated by pollutants and contaminants. Masonry is permeable to wind driven

[1]Consulting architectural engineer and member of the architectural engineering faculty, University of Texas at Austin, Austin, TX 78758.

[2]For the readers convenience the italic numbers in brackets refer to citations found in the Bibliography appended to this paper.

rain, which may contain pollutants, and masonry may contain contaminants. Therefore, metals in masonry walls must be properly protected to preserve structural integrity. Connectors are usually galvanized. Lintels and shelf angles are usually protected by paint and flashing, but they may also be galvanized.

Corrosion

Corrosion is the chemical or electrochemical oxidation of the surface of metal which can result in loss of material or accumulation of deposits [*18*]. The electrochemical basis for localized corrosion is described by Pourbaix [*49*] and others [*21*]. Corrosion of carbon and low-alloy steels is primarily governed, in most cases, by the combined action of water and oxygen [*36*]. Corrosion increases with increased quantities of water, oxygen, pollutants, and contaminants reaching the steel surface [*14,36,45,58*]. A corrosive environment may be any form of moisture ranging from plain water to the strongest acids or alkali [*15*].

Water Permeance

Masonry is water permeable. A 4-in. (100 mm) wythe of well built brick masonry is not a barrier to wind driven rain [*22,27,38,39,54,59,67,68*]. These published data are confirmed by several field tests directed by this author on masonry in situ on buildings, using an adaptation of ASTM Test Method for Water Permeance of Masonry (E 514). Stockbridge [*62*] has said, "Today it is an accepted fact that no matter how good the workmanship and how impervious the units and mortar used, water will penetrate exterior masonry walls." The quantity of water entering a brick wall is related to the driving rain index; the quality of mortar joint tooling; the porosity of the masonry units; the size, number, and extent of cracks in the face of the masonry units; and the size, number, and extent of structural cracks in the wall.

For example, in Cleveland, Ohio, building walls have a severe exposure to driving rain, while walls in Phoenix, Arizona, may have a slight exposure [*30*]. Cracks in brick increase water permeance [*47*]. Wind driven rain penetrates cracks wider than 0.1 mm (0.0039 in.) [*1*]. Improperly tooled mortar joints leave cracks between brick and mortar which admit wind-driven rain [*31*]. DeBruyn [*17*] and others have discussed the importance of cracks in permitting ready access of corrosives to metals and in initiating corrosion. Every effort should be made to keep the size and occurrence of cracks to a minimum in a structure exposed to a corrosive environment [*45*].

Masonry walls are subject to surface and interstitial condensation [*41*]. Condensation typically occurs on the interior face of the exterior brick masonry wythe [*42*]. This water drains to and accumulates on lintels and shelf angles unprotected by flashing. However, flashing does not prevent surface condensation on shelf angles and lintels, which are located very near exterior air and which have high thermal conductivity. Water may also enter the wall through roof leaks and window sills and at wall projections and recesses [*24*], as well as through sealant joints [*31*].

The quantity of water permeating through the masonry is related to the solidity of mortar joints [*31*]. Large voids in mortar or grout adjacent to steel accelerate corrosion [*45*]. Full mortar joints are required by several building codes [*6, 7, 60, 65*]. Water permeance is also related to bond strength of mortar to brick [*53*]. Four-inch (100 mm) brick masonry walls built of mortar containing vinylidene chloride latex additive, like all other 4-in. (100 mm) brick walls, are permeable to water, but walls containing such additives are typically less permeable than those built without the additive [*38*].

Metals begin to corrode at an accelerated rate when the relative humidity of the air layer next to the surface exceeds 75% [*56*]. The corrosion of steel has been related to the percentage time-of-wetness, that is, the percentage of the total time that the relative humidity exceeds a critical value [*56*]. The time-of-wetness in the vent space of an exterior wall has been found to average about the same as that on the exterior [*56*]. For example, the relative humidity in Cleveland, Ohio, exceeds 75% for about 38% of the time. The relative humidity of mortar in masonry walls exceeds 75% for extended periods of time [*33*]. It is well established that the exterior wythe of brick masonry may be damp for significant periods [*44*].

Pollutants

Atmospheric pollutants, which contribute to corrosion, include sulfur dioxide (SO_2), hydrogen sulphide (H_2S), ammonia (NH_3), nitrite (NO_2), nitrate (NO_3), sodium chloride (NaCl), and particulate matter [*57*]. Sulfur dioxide in air combines with rain and oxygen to yield sulfuric acid. The pH of the annual precipitation, for example, in Cleveland, Ohio, was 4.3 in 1972–1973. The ill effects of acid rain on masonry was recognized at least as early as 1931. One molecule of sulfur dioxide may lead to 15 to 40 molecules of rust [*21*]. The SO_2 content of air varies greatly among and within cities. For example, between December 1963 and July 1965, the average SO_2 content of air in Cleveland, Ohio, was 89% greater than that in South Bend, Pa., and the corrosion rate of steel was found to be 8% to 41% greater in Cleveland than in South Bend [*16*]. In another study (1960–62), the corrosiveness to steel of the atmosphere in Cleveland, Ohio, was 170% greater than in Detroit, Mich., 73% greater than in Waterbury, Conn., and 28% greater than in Pittsburgh, PA. [*16*]. Corrosive effects of air pollutants may vary 300% within a single city [*58*].

Contaminants

Contaminants in the steel or masonry materials may accelerate corrosion. Impurities in steel, differences in grain boundaries, or surface texture may foster corrosion [*45*]. Salts in masonry tend to increase corrosion rate due to a combination of increased electrolytic conductivity and the formation of non-protective corrosion products [*21,45,64*]. The various ions present in masonry which could influence corrosion include hydroxyl, calcium, sodium, potassium, carbonate, sulfate, and chloride ions [*45*].

Brick may contain chlorides, calcium sulfate, vanadyl sulfate, or manganese oxide [*20*]. Mortar may contain calcium carbonate, sodium carbonate, or potassium carbonate. Cement may contain calcium hydroxide [*20*] or chlorides [*66*]. Mortar aggregates may contain chlorides [*28*]. Cement may react with brick to yield sodium sulfate or potassium sulfate [*20*]. Acids used for cleaning walls may contain potassium chloride, vanadyl chloride, hydrogen chloride, or organic acids. Mortar mixing water and additives may contain chloride, for example, vinylidene chloride latex. Calcium chloride may be added to mortar in winter to accelerate mortar set. However, some of the chloride becomes intimately bound with the hydrated cement phases in mortar and is not easily soluble in water [*50*]. It is generally agreed that the amount of calcium chloride in flake form in concrete should not exceed 2% of the cement by weight [*50*]. The American Concrete Institute has suggested that chloride ion concentrations exceeding 400 to 500 ppm in concrete might be considered dangerous and that levels well below these values should be maintained, if practicable [*11*]. However, the concentration of chloride in corrosion products collected from steel samples exposed to the atmosphere (that is, not in contact with masonry) for one year (1961-1962) in Cleveland, Ohio, was 447 ppm [*57*].

Corrosion may also occur as a result of stray direct electrical currents from elevators or computers, or from contact by different metals [*28*].

If sealant joints at shelf angles are placed after the wall has been cleaned with acid, the shelf angle is exposed to an acid bath, which may very often be a strong acid solution. If installation of the sealant joint at the shelf angle is long delayed, perhaps for as much as two years, the steel is exposed to cascades of acid rain, as water runs down the wall and is forced by wind into open joints.

Performance

The fact that water leaking through brick masonry walls causes corrosion of steel in buildings has been recognized for more than 30 years [*5*]. Severe corrosion of steel in reinforced masonry has occurred in seven or eight years but more usually in 20 to 25 years [*52*]. Since laboratory accelerated tests designed to predict atmospheric corrosion performance for low-alloy steels have for the most part been ineffective [*36*], investigators of corrosion in masonry have examined buildings in situ or have exposed masonry test specimens with embedded steel to natural environments.

Fishburn [*23*] noted that galvanized, corrugated steel ties in masonry wall specimens exposed to weather begun to rust after 180 days of exposure. Kropp and Hilsdorf [*69*] found corrosion of plain and galvanized steel in mortar joints after six months of exposure at 90% relative humidity. Brand [*3*] found corrosion damage on many metal ties in masonry, ranging from surface rust to complete penetration of the metal section. Foster and Thomas [*4*] observed corrosion on uncoated reinforcing bars placed vertically in the cores of brick and placed horizontally in mortar joints in brick wall specimens exposed to weather for two years. Galvanized rods in the same exposure provided complete steel protection. In the same test, galvanized ladder type joint reinforcement had some heavy pitting.

Moore [*44*] examined the galvanized steel ties in cavity walls in 186 buildings and found a corrosion rate of 10 to 20 g/m^2/year (0.033 to 0.066 oz psf/yr). Based on those data, zinc coatings on steel used for ties and anchors in the U.S. have the life expectancies shown in Table 1.

In 1979, Dr. Lewis W. Gleekman, now deceased and then president of Materials and Chemical Engineering Services, Southfield, Mich., wrote to this author, expressing the opinion that galvanized steel anchors typically used to anchor brick veneer to steel studs have an estimated life expectancy ranging from seven years under the worst conditions to more than 20 years under ideal conditions.

During 1980 and 1981 this author examined the condition of steel lintels and shelf angles in conventional masonry on 18 buildings on which there was some exterior indication of possible corrosion. Eleven of the buildings were in the area of Cleveland, OH, and seven were in the area of Detroit, MI. The buildings were 4 to 26 years old. Data from that survey are presented in Table 2. Photographs of some elements are presented in Figs. 1 through 8. Corrosion was judged to be severe in 12 of the 18 buildings and on 8 of the 12 buildings that were ten years old or less.

Protection

Zinc (galvanized) coatings protect steel in two ways, i.e., as a barrier separating steel from water and oxygen and as a sacrificial anode at uncoated areas, for example, cut edges and scratches [*58*]. Effectiveness of zinc coatings depend on their thickness and the environment [*58*]. DeVekey [*70*] has recommended austenitic stainless steel (18% + chromium and 8% nickel) for severe exposure conditions.

Corrosion resistant metals for ties and anchors in masonry are required by law and recommended good practice (*2, 3, 7, 46, 60, 65*]. Of those codes only BIA [*7*] and ACI [*6*] require a specific quality of corrosion resistance material. The BIA code requires 0.8 oz psf (244 g/m^2) zinc galvanized wire (ASTM A116-66, Class 3) or equivalent for ties, and 1 to 1.5 oz psf (305 to 458 g/m^2) hot dip galvanizing (ASTM A153-66) for anchors. The ACI 531-79 code requires 0.3 to 0.4 oz psf (122 g/m^2) (ASTM A116, Class 1) zinc galvanizing for wire ties entirely embedded in mortar or grout; 0.8 oz psf (244 g/m^2) (ASTM A116, Class 3) galvanizing for wire ties when not entirely embedded in mortar or grout; and 1 to 1.5 oz psf (305 to 458 g/m^2) galvanizing for anchors or equivalent. Based on

TABLE 1—*Zinc coatings on steel ties and anchors in masonry.*

Element	Size		ASTM Specification Reference and Class	Minimum Coating, oz psf	Average Weight, g/m²	Life Expectancy of Coating, year		
	No.	in. (mm)				Min[c]	Mode	Max
Wire ties	10 through 12	0.105 (2.69) to 0.135 (3.43)[a]	A 116 Class 1 or A 641 Class 1	0.3	99	5	7	9
Wire ties	9 and larger	0.144 (3.76)[a] and over	A 116 Class 1 or A 641 Class 1	0.4	122	6	9	12
Wire ties	7 through 12	0.105 (2.69) to 0.177 (4.5)[a]	A 116 Class 3 or A 641 Class 3	0.8	244	12	18	24
Anchor bolt	...	0.375 (9.52)[a] and under	A 153 Class D	1.0	305	13	22	30
Anchor bolt	...	Over 0.375 (9.52)[a]	A 153 Class C	1.25	381	15	27	38
Sheet metal anchors	...	8 (203)[b] and under	A 153 Class B-3	1.3	397	17	28	39
Sheet metal anchors	...	over 8 (203)[b]	A 153 Class B-2	1.5	458	19	32	46
Reinforcing rods	...	0.25 (6.35)[b]	A 123	2.3	702	30	50	70

[a]Diameter.
[b]Length.
[c]Based on corrosion rate of 0.033 to 0.066 oz psf-year [*44*] and minimum coating requirement for individual specimen to nearest whole year.

TABLE 2—*Survey of corrosion on shelf angles and lintels supporting masonry.*

Building number	1	2		3
Building name	Building 2255 Pine Ridge Apts.	J. M. Gallager Jr. High School		Watterson Lake Jr. High School
Building address	I90 & Rt. A 84 Cleveland, OH	Franklin at W. 65 Cleveland, OH		1422 W. 74th St. Cleveland, OH
Approximate time of building completion	1971	1977		1970
Time of building observation	June 1981	June 1981		June 1981
Approximate age of building, years	10	4		11
Object	lintel	shelf angle		shelf angle
Environment:				
wall orientation	W	S	N	E
story	1	2	1	1
steel coating	paint	paint	paint	paint
flashing	yes	yes	yes	yes
weep holes	none	none	none	none
Degree of corrosion	slight	severe	severe	slight to severe
Figure no.	none	1	none	2

Building number	4	5	6
Building name	East Tech High School	Benesch Elementary School	East High School
Building address	2439 E. 55th Cleveland, OH	5393 Quincy Ave. Cleveland, OH	1349 E. 79th St. Cleveland, OH
Approximate time of building completion	1972	1975	1976
Time of building observation	April 1981	April 1981	April 1981
Approximate age of building, years	9	6	5
Object	shelf angle	shelf angle	shelf angle
Environment:			
wall orientation	S	S	W
story	3	1	1
steel coating	paint	paint	paint
flashing	yes	yes	none
weep holes	none	some	none
Degree of corrosion	moderate to severe	slight to moderate	moderate
Figure no.	none	none	none

Building number	7	8	9
Building name	Cedar Apts.	Wade Park Elementary School	Sears Tower Case-Western Medical Center
Building address	E. 30 & Cedar Rd. Cleveland, OH	7600 Wade Park Cleveland, OH	10900 Euclid Ave. Cleveland, OH
Approximate time of building completion	1954	1975	1967
Time of building observation	Aug. 1980	Aug. 1981	Aug. 1981
Approximate age of building, years	26	6	14
Object	lintel	lintel	shelf angle

TABLE 2—*Continued.*

Building number	**7**	**8**	**9**	
Environment:				
wall orientation	W		N	W
story	2	1	10	9
steel coating	paint	paint	paint	paint
flashing	none	yes	yes	yes
weep holes	none		none	none
Degree of corrosion	slight to moderate	severe	severe	severe
Figure no.	none	none	none	3
Building number	**10**	**11**	**12**	
Building name	Auburn Science and Engineering Center	Central Trust of NE OH	Jewish Fed. Apts.	
Building address	Akron, OH	Court Ave., SW & Tuscarwas St. W. Canton, OH	10 Mile Road Detroit, MI	
Approximate time of building completion	1967	1972	1972	
Time of building observation	Aug. 1980	Aug. 1980	June 1981	
Approximate age of building, years	13	8	9	
Object	shelf angle	shelf angle	tie	
Environment:				
wall orientation	NW	W	W	
story	stair tower	13	15	
steel coating	paint	paint	galvanized	
flashing	yes	yes	NA[a]	
weep holes	none	none	NA	
Degree of corrosion	severe	moderate to severe	severe	
Figure no.	4	5	none	
Building number	**13**		**14**	
Building name	Parking Garage Wayne State University		Travelers Building	
Building address	450 Palmer St. Detroit, MI		26555 Evergreen St. Southfield, MI	
Approximate time of building completion	1967		1971	
Time of building observation	June 1981		July 1981	
Approximate age of building, years	14		10	
Object	shelf angle	lintel	tie	columns
Environment:				
wall orientation	S	E	NE corner	NE corner
story	2	1	1	1
steel coating	paint	paint	galvanized	none
flashing	yes	none	NA	NA
weep holes	none	none	NA	none
Degree of corrosion	severe	severe	severe	moderate
Figure no.	6	none	7	none

TABLE 2—*Continued.*

Building number	**15**	**16**	**17**
Building name	Howard Johnson Motel	Kiwanian Home	Handicap Recreation Center
Building address	Mich. & Wash. Ave. Detroit, MI	1270 Electric St. Lincoln Park, MI	100 Lennox St. Detroit, MI
Approximate time of building completion	1966	1975	1967
Time of building observation	July 1981	Aug. 1981	Aug. 1981
Approximate age of building, years	15	6	14
Object	shelf angle	plate	lintel
Environment:			
wall orientation	E	N	N
story	14	1	1
steel coating	none?	paint	paint
flashing	none	none	none
weep holes	none?	none	none
Degree of corrosion	severe	severe	severe
Figure no.	none	none	none
Building number	**18**		
Building name	Keller & Pollic Bldg.		
Building address	23366 Commerce Park S. Beachwood, OH		
Approximate time of building completion	1971		
Time of building observation	April 1981		
Approximate age of building, years	10		
Object	lintel		
Environment:			
wall orientation	S		
story	1		
steel coating	paint		
flashing	yes		
weep holes	none		
Degree of corrosion	severe		
Figure no.	8		

[a]NA = not applicable.

findings by Moore [*44*], the life expectancy under the BIA code [*7*] is about 16 years for ties and 20 to 30 years for anchors. Under the ACI code [*6*] the life expectancy is about 7 years for embedded ties and 16 years for ties not embedded.

Protection from corrosion of reinforcing steel in brick masonry is provided for by requiring solid embedment of steel in mortar or grout and a specified minimum thickness of masonry between any bar and the nearest exterior face of the masonry [*2,6,7,60,65*]. The cover requirement in walls exposed to weather is 50 mm (2 in.) for bars larger than No. 5 and 38 mm (1½ in.) for No. 5 bars or smaller. Reinforcement 6 mm (¼ in.) or less in diameter embedded in

FIG. 1—*Corroded lintel.*

FIG. 2—*Corroded lintel.*

FIG. 3—*Corroded lintel.*

FIG. 4—*Corroded lintel.*

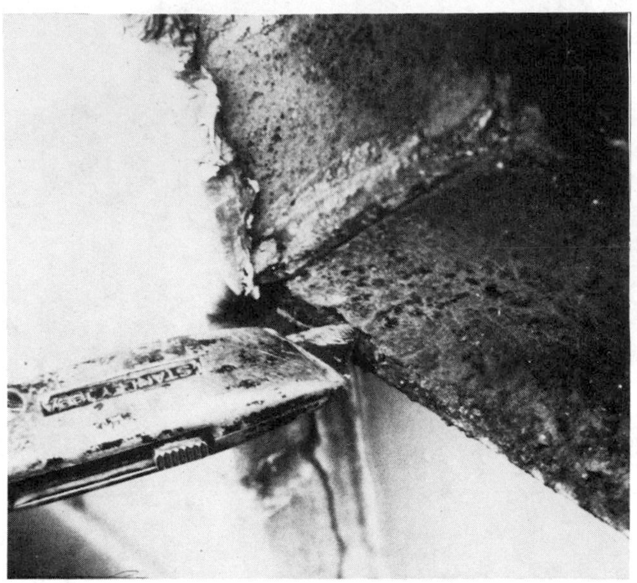

FIG. 5—*Corroded lintel.*

FIG. 6—*Corroded shelf angle.*

FIG. 7—*Corroded tie.*

horizontal mortar joints is required to have a mortar cover of not less than 16 mm (5/8 in.) [*7*]. These are the same cover requirements as for concrete [*8*]. However, because masonry is less dense and more permeable than structural concrete, it is reasonable to expect that more protection of steel would be required in brick masonry than in concrete. Indeed, some authorities have so recommended [*19,25*]. They suggest that reinforcing steel in brick masonry be provided with at least 4 in. (100 mm) of cover or that the steel be galvanized.

Steel reinforcing bars have been placed in the cores of solid brick masonry units. A solid masonry unit may be cored to 25% of the gross cross-sectional area [ASTM Specification for building brick (C 62) and Specification for Facing Brick (C 216)]. The maximum width of a core is 35 mm (1 3/8 in.) in a brick 92 mm (3 5/8 in.) in width. The minimum thickness of grout or mortar between brick and reinforcement is 6.35 mm (1/2 in.) [*7*]. To provide adequate grout or mortar thickness for a reinforcing bar in the core of a brick requires a precision of bar placement which is impractical and virtually impossible with normal and customary masonry construction techniques. "Slushing" mortar into small spaces does not fill the space solidly [*40*]. To provide a No. 3 bar with the minimum required cover of 38 mm (1 1/2 in.) in the core of a 92 mm (3 5/8 in.) wide brick would require perfect placement of the bar in the exact center of the core, which precision is unattainable.

For masonry exposed to driving rain and freezing while wet, subject to condensation or corrosive fumes, or exposed to alternate wetting and drying, the

FIG. 8—*Corroded lintel.*

current draft of the British Standard BS 5628, Code of Practice for the Structural Use of Masonry, Part 2, Reinforced and Prestressed Masonry, requires a cover to the exterior face of the masonry of 2 in. (50 mm) of mortar or grout plus the shell thickness of the masonry unit, except that the cover may be reduced to 32 mm (1¼ in.) of mortar or grout plus the shell thickness of the masonry unit, if the steel is galvanized with a zinc coating of 1 oz psf (305 g/m^2). The same draft standard requires that bed joint reinforcement be galvanized and that 32 mm (1¼ in.) of cover be provided. However, these requirements have been attacked as too conservative for brick masonry by Foster and Thomas [*26*], but have been found acceptable by Roberts [*55*] for concrete block masonry.

Corrosion of structural steel lintels and shelf angles may be reduced by keeping the steel as dry as possible with flashing and paint or by galvanizing. The necessity for protecting structural steel members, supporting exterior masonry, from water which has permeated masonry due to wind driven rain has been recognized for at least the last half century [*35*]. The Steel Structures Painting Council currently recommends that steel enclosed in masonry in weathertight building frames not subject to condensation, in dry void spaces or below 70% relative humidity, may be protected from corrosion by one shop coat of paint [*34*]. Obviously this does not apply to shelf angles and lintels with or without flashing. The presence of a shop coat of paint on structural steel is insufficient to prevent severe corrosion caused by water leakage [*61*]. Where structural steel is frequently but not continuously wet by fresh water, for example, condensation, the Council recommends either: (*a*) blast cleaning or pickling with either three coats of alkyd, phenolic, or vinyl paint, or two coats of epoxy-polyamide paint, or one coat (two to five mil dry thickness) of zinc-rich paint; or (*b*) wire brushing and four coats of oil base paint [*34*]. The corrosion rate on improperly coated metal is often far more rapid and more destructive than on uncoated metal [*71*].

Flashing may be used to keep water that permeates or condenses on the masonry from reaching spandrel beams, shelf angles, and lintels. Spandrel beams should be protected by impermeable membranes [*31*]. Improperly built spandrel beam "waterproofing" can increase water permeance [*12*]. The practice of keeping the outside edge of flashing behind the exterior wall surface 13 mm (½ in.) for aesthetic reasons invites trouble [*43*]. If galvanized steel is not used for lintels and shelf angles, continuous flashing should be installed except where the annual precipitation is less than 760 mm (30 in.) [*10,63*]. This author made a search of the architectural and engineering literature in the English language in the U.S. from 1937 through 1979 to determine design recommendations for flashing of spandrel beams and shelf angles supporting brick masonry exterior walls. Seventeen references were found, and all recommended the use of flashing at such locations, except in arid regions or unless galvanized structural steel is used.

Flashing, when used, should be placed immediately above the horizontal leg of structural steel supporting masonry exterior walls. Flashing should extend beyond the exterior wall face and turned down at 45 deg to form a drip projecting from the wall no less than 19 mm (¾ in.), and extend on the interior up over the steel but not less than 6 to 9 in. (150 to 230 mm) and into a reglet, interior masonry wythe, or behind sheathing. Flashing joints should be completely sealed and provided with sealed end dams, that is, turned up 6 to 9 in. (150 to 230 mm) at abutting columns and at expansion or control joints in the masonry [*31*].

Weep holes at least 6 mm (¼ in.) in diameter, extending at least 75 mm (3 in.) into the masonry, should be located at least 610 mm (24 in.) on centers horizontally immediately above the horizontal leg of all internal or cap flashing and above galvanized lintels or shelf angles [*31*]. Weep holes in masonry at lintels, shelf angles, and at the bottom course of masonry veneer are required by law [*60*] and recommended as good practice [*31*].

A horizontal expansion joint should be installed immediately below shelf angles, which necessitates a sealant joint at the wall face immediately below and in front of the shelf angle toe [*4*]. The life expectancy of sealants is four to nine years [*13*]. In the absence of proper sealant joint installation and maintenance, shelf angles are vulnerable to acid rain penetration. If the flashing extends beyond the wall face and is turned down, the sealant and its backer rod can be placed in a more protected position in the dihedral angle between the underside of the extended flashing and the face of the masonry. Alternatively, if stiff, for example, metal, flashing is used with its outer edge flush with the face of the wall, the sealant joint can also be placed immediately below the flashing and flush with the wall face [*48*].

Conclusions

Metals frequently placed in masonry include ties, anchors, joint reinforcement, structural reinforcing rods, and structural steel members. Steel in masonry must be protected from corrosion. A nominal 100-mm (4-in.) thick brick masonry wythe is permeable to wind-driven rain. Water may also enter through cracks in units or walls, condensation, roof and sill leaks, and sealant joints. Good workmanship, for example, well-tooled and filled mortar joints, properly installed flashing and weep holes, and properly built and maintained sealant joints, reduce water permeance and corrosion but do not eliminate it. Atmospheric pollutants and masonry contaminants contribute to metal corrosion in masonry, the occurrence of which is well documented by field surveys of masonry in situ.

Protective measures for steel in masonry include galvanizing for ties, anchors, and fasteners; cover alone or galvanizing and cover for reinforcing steel; and flashing with weep holes and paint or galvanizing for structural steel with properly installed sealant joints at shelf angles.

Recommendations

1. Building code requirements and construction specifications for galvanized steel in exterior masonry for ties, anchors, fasteners, and joint reinforcing; for reinforcing bars with less than 4 in. (100 mm) of masonry cover; and for structural steel lintels and shelf angles and their washers and bolts without flashing should be hot-dip zinc coated in conformance with the requirements of ASTM Specification for Zinc Coating (Hot-Dip) on Iron and Steel Hardware (A 153), which would provide a life expectancy of 15 to 60 years, depending on item thickness and length. For ties, anchors, and joint reinforcement, copper-coated wire (ASTM B 227, grade 30 HS) may be used in lieu of hot dip galvanizing.

2. Unless construction operations are professionally observed to ensure proper application of paint and installation of flashing and weep holes, structural steel lintels and shelf angles and their washers and bolts should be hot dip galvanized. Four coats of oil paint or three coats of alkyd, phenolic, or vinyl paint or two coats of epoxy-polyamide paint, or one coat of zinc-rich paint on an appropriately cleaned surface may be used in lieu of hot dip galvanizing.

Acknowledgments

The author is indebted to many clients, students, and colleagues who have contributed much to his knowledge of masonry over the years. This work was supported by a grant from the Dow Chemical Company.

Bibliography

[1] Birkeland, O. and Sevendsen, S. D., "Norwegian Test Methods for Rain Penetration through Masonry Walls," *Symposium on Masonry Testing, ASTM STP 320*, American Society for Testing and Materials, Philadelphia, 1963, p. 3.

[2] *BOCA Basic Building Code*, Building Officials & Code Administrators International, Inc., Chicago, 1978.

[3] Brand, R. G., "High-Humidity Buildings in Cold Climates—A Case History," *Durability of Building Materials and Components, ASTM STP 691*, American Society for Testing and Materials, Philadelphia, 1980, pp. 231-238.

[4] "Brick Masonry Cavity Walls," Technical Notes on Brick Construction, No. 21, Brick Institute of America, McLean, VA, Jan./Feb. 1977 (also No. 21A, May/June 1977, and Aug./Sept. 1965).

[5] Brown, B., "Leaky Walls Endanger Steel Buildings," *Engineering News Record*, New York, Dec. 8, 1949, p. 42.

[6] *Building Code Requirements for Concrete Masonry Structures*, ACI 131R-79, American Concrete Institute, Detroit, MI, 1978.

[7] *Building Code Requirements for Engineered Brick Masonry*, Brick Institute of America, McLean, VA, 1969.

[8] *Building Code Requirements for Reinforced Concrete*, ACI 318-77, American Concrete Institute, Detroit, MI, 1977.

[9] Cady, P. D., "Corrosion of Reinforcing Steel," *Significance of Tests and Properties of Concrete and Concrete-Making Materials, STP 169B*, American Society for Testing and Materials, Philadelphia, 1978, pp. 275-299.

[10] "Cavity Walls in Skeleton Frame," *Technical Notes on Brick and Tile Construction*, Vol. 10, No. 5, Brick Institute of America, McLean, VA, May 1959.

[11] *Commentary on Building Code Requirements for Reinforced Concrete*, ACI 318-77, American Concrete Institute, Detroit, MI, Dec. 1977, Sect. 3.4.2.

[12] Connor, C. C., "Factors in the Resistance of Brick Masonry Walls to Moisture Penetration," *Proceedings of the American Society for Testing and Materials*, Vol. 48, Philadelphia, 1948.

[13] Cook, J. P., *Construction Sealants and Adhesives*, Wiley Interscience, New York, 1970, p. 54.

[14] Copson, H. R., "A Theory of the Mechanism of Rusting of Low-Alloy Steel in the Atmosphere," *Proceedings of the American Society for Testing and Materials*, Vol. 45, Philadelphia, 1945, p. 554.

[15] *Corrosion in Action*, The International Nickel Company, Inc., New York, 1977, p. 6.

[16] "Corrosiveness of Various Atmospheric Test Sites as Measured by Specimens of Steel and Zinc," *Metal Corrosion in the Atmosphere, ASTM STP 435*, American Society of Testing and Materials, Philadelphia, 1968, pp. 360-391.

[17] deBruyn and Lobry, C. A., "Influence of Bond and Cracking on Corrosion of Reinforcement," RILM, Symposium on Bond and Crack Formation in Reinforced Concrete, Stockholm, 1957, *Summaries*, 4, 17-19, 1959.

[18] "Definition of Terms Relating to Metallography," *Annual Book of ASTM Standards*, American Society for Testing and Materials, Philadelphia, 1976.

[19] *Design Guide for Reinforced and Prestressed Clay Brickwork*, British Ceramic Research Association, Stoke-on-Trent, England, 1977.

[20] "Effloresence, Prevention and Control," *Technical Notes on Brick and Tile Construction*, No. 23A, Brick Institute of America, McLean, VA, Jan. 1970.

[21] Evans, V. R. and Taylor, C. A. J., "Mechanism of Atmospheric Rusting," *Corrosion Science*, Vol. 12, Pergamon Press, London, 1972, pp. 227-246.

[22] Fishburn, C. C., Parsons, D. E., and Peterson, P. H., *Effect of Outdoor Exposure on the Water Permeance of Masonry Walls*, BMS, No. 76, National Bureau of Standards, U.S. Government Printing Office, Washington, DC, Aug. 15, 1941.

[23] Fishburn, C. C., *Strength and Resistance to Corrosion of Ties for Cavity Walls*, Report BMS 101, National Bureau of Standards, U.S. Government Printing Office, Washington, DC, July 1, 1943.

[24] "Flashing Clay Masonry," Technical Notes on Brick and Tile Construction, No. 7A, Brick Institute of America, McLean, VA, Feb. 1962.

[25] Foster, D. and Thomas, A., "Some Interim Comments on the Corrosion of Reinforcement in Brickwork," *Proceedings of the British Ceramic Society*, Stoke-on-Trent, England, Sept. 1975, pp. 197-223.

[26] Foster, D. and Thomas, A., "Aspects of Durability of Clay Brickwork," *Institution of Structural Engineers Symposium*, ISE, London, July 1981.

[27] Fowler, D. W. and Grimm, C. T., "Effect of Sandblasting and Face Grouting on Water Permeance of Brick Masonry," *Masonry Past and Present, ASTM STP 589*, American Society for Testing and Materials, Philadelphia, 1975, pp. 225-271.

[28] *Galvanized Reinforcement for Concrete*, International Lead Zinc Research Organization, Inc.

[29] Grimm, C. T., "Metal Ties and Anchors for Brick Walls," *Journal of the Structural Division*, ASCE, Vol. 102, No. ST4, April 1976, pp. 339-858.

[30] Grimm, C. T., "Driving Rain Index for Masonry Walls," *Symposium on Masonry: Materials, Properties and Performance*, American Society for Testing and Materials, Philadelphia, Dec. 9, 1980 (in publication).

[31] Grimm, C. T., "Water Permeance of Masonry Walls—A Review of the Literature," *Symposium on Masonry: Materials, Properties and Performance*, American Society for Testing and Materials, Philadelphia, Dec. 9, 1980 (in publication).

[32] Houston, J. T., Atimtay, E., and Ferguson, P. A., *Corrosion of Reinforcing Steel Embedded in Structural Concrete*, Center for Highway Research, University of Texas at Austin, Austin, TX, March 1972.

[33] Isberner, A. W., "Properties of Masonry Cement Mortars," *Designing, Engineering and Constructing with Masonry Products*, Gulf Publishing Co., Houston, TX, 1969, pp. 42-50.

[34] Keane, J. D., Ed., *Steel Structures Painting Manual*, Vol. 2, Systems and Specifications, Steel Structures Painting Council, Pittsburgh, PA, 1973, p. 63.

[35] Kidder, F. E., and Parker, H., *Architects' and Builders' Handbook*, John Wiley and Sons, Inc., New York, 1931, pp. 946-947.

[36] Leckie, H. P., "Corrosion Standards and Control in the Iron and Steel Industry," *Manual of Industrial Corrosion Standards and Control, ASTM STP 534*, Philadelphia, 1973, p. 212.

[37] *Manual of Steel Construction*, American Institute of Steel Construction, New York, 1970, pp. 5-311.

[38] *Masonry Construction With High Bond Mortar*, Dow Chemical Company, Midland, MI, Aug. 1967, p. 10.

[39] McBurney, J. W., Copeland, M. A., and Brink, R. C., "Permeability of Brick-Mortar Assemblages," *Proceedings*, American Society for Testing and Materials, Philadelphia, 1946, pp. 1333-1354.

[40] "Moisture Control in Brick and Tile Walls—Rain Penetration," Technical Notes on Brick and Tile Construction, No. 7B, Brick Institute of America, McLean, VA, Jan. 1965, p. 4.

[41] "Moisture Control in Brick and Tile Walls—Condensation," Technical Notes on Brick and Tile Construction, No. 7C, Brick Institute of America, McLean, VA, Feb. 1965.

[42] "Moisture Control in Brick and Tile Walls—Condensation Analysis," Technical Notes on Brick and Tile Construction, No. 7D, Brick Institute of America, McLean, VA, June 1978.

[43] Monk, Jr., C. B., "Masonry Facade and Paving Failures," *Proceedings of the Second Canadian Masonry Symposium*, Carleton University, Ottawa, Ontario, Canada, June 9-11, 1980, pp. 469-480.

[44] Moore, J. F. A., "The Performance of Cavity Wall Ties," *Building Research Establishment Information Paper*, IP 4/81, Building Research Station, Watford, England, April 1981.

[45] Mozer, J. D., Bianchini, A. C., and Kesler, C. K., "Corrosion of Reinforcing Bars in Concrete," *Journal of American Concrete Institute*, Detroit, MI, Aug. 1965, pp. 909-930.

[46] *National Building Code*, American Insurance Association, New York, 1967.

[47] Palmer, L. A. and Parsons, D. A., "Permeability Tests on Eight Inch Brick Wallettes," *Proceedings of the American Society for Testing and Materials*, Part II, Vol. 34, Philadelphia, 1934, pp. 419-453.

[48] Parise, C. J., "Shelf Angle Components in Cavity Wall Construction," *Symposium on Masonry Materials, Properties and Performance*, American Society for Testing and Materials, Philadelphia, (in publication).

[49] Pourbaix, M., *Localized Corrosion*, National Association of Corrosion Engineers, Houston, TX, 1974, pp. 12-26.

[50] Ramachandran, V. S., "Calcium Chloride in Concrete," *Canadian Building Digest*, No. 165, National Research Council of Canada, Ottawa, Ontario, Canada, 1974.

[51] "Reinforced Brick and Tile Lintels," Technical Notes on Brick and Tile Construction, No. 17H, Brick Institute of America, McLean, VA, Nov.-Dec. 1964.

[52] Rengaswami, N. S., Balesubramanyam, T. M., Ventkataraman, H. S., and Rajagopalan, K. S., "Corrosion of Reinforcement in Reinforced Concrete and Reinforced Brickwork," *Indian Concrete Journal*, June 1964, pp. 233-237.

[53] Ritchie, T. and Davidson, J. I. "Factors Affecting Bond Strength and Resistance to Moisture Penetration of Brick Masonry," *Symposium on Masonry Testing, ASTM STP 320*, American Society for Testing and Materials, Philadelphia, Feb. 1963, pp. 16-30.

[54] Ritchie, T., "Water Permeance Tests of TTW Brick Walls," Building Research Note, No. 86, National Research Council, Ottawa, Ontario, Canada, May 1972.

[55] Roberts, J. J., "Interim Results from an Investigation of the Durability of Reinforcing Steel in Reinforced Concrete Blockwork," *Institution of Structural Engineers Symposium*, ISE, London, July 1981.

[56] Sereda, P. J., "Atmospheric Corrosion of Metals," *Canadian Building Digest*, CBD 170, National Research Council of Canada, Ottawa, Ontario, Canada, 1975.

[57] Sereda, P. J., "Weather Factors Affecting Corrosion of Metals," *Corrosion in Natural Environments, STP 558*, American Society for Testing and Materials, Philadelphia, 1974, pp. 7-22.

[58] Shaffer, T. F., "Specifying Galvanized Steel for Construction Products," *The Construction Specifier*, CSI, Alexandria, VA, Oct. 1973.

[59] Skeen, J. W., "Experiments on Rain Penetration of Brickwork—The Effect of Mortar Type," *Transactions and Journal of the British Ceramic Society*, Vol. 70, No. 1, Jan. 1971, pp. 27-30.

[60] *Standard Building Code*, Southern Building Code Congress International, Inc., Birmingham, AL, 1979.

[61] Stetina, H. J., "Structural-Steel Construction," *Building Construction Handbook*, McGraw-Hill, New York, 1975, pp. 6-88.

[62] Stockbridge, J. G., "Evaluation of Terra Cotta on In-Service Structures," *Durability of Building Materials and Components, ASTM STP 691*, American Society for Testing and Materials, Philadelphia, 1980, pp. 216-230.

[63] "Structural Steel Lintels," Technical Notes on Brick and Tile Construction, No. 31B, Structural Clay Products Institute (Brick Institute of America), McLean, VA, Sept. 1969.

[64] Uhlig, H. H., *Corrosion and Corrosion Control*, Wiley, New York, 1971.

[65] *Uniform Building Code*, International Conference of Building Officials, Whittier, CA, 1976.

[66] Verbeck, G. J., "Mechanisms of Corrosion of Steel in Concrete," *Corrosion of Metals in Concrete*, ACI SP-49, American Concrete Institute, Detroit, MI, 1975, pp. 21-38.

[67] West, H. W. H., Ford, R. W., and Peake, F., "Single Leaf Masonry: The Resistance to Rain Penetration of Some Composite Systems," *Proceedings of Third International Brick Masonry Conference*, Bundersverband der Deutschen Ziegslindustrie, Bonn, West Germany, 1975, pp. 462-468.

[68] West, H. W. H., Ford, R. W., and Peake, F., *The Effect of Suction Rate of Bricks on Permeability of Mortar*, Technical Note No. 254, British Ceramic Research Association, Stoke-on-Trent, England, March 1976.

[69] Kropp, J. and Hilsdorf, H. K., "Evaluation of Corrosion Resistance of Reinforcement Embedded in Masonry Joints," *Proceedings of the Fifth International Brick Masonry Conference*, Brick Institute of American, McLean, VA, Oct., 1980 (in publication).

[70] De Vekey, R. C., "Durability of Reinforced Masonry," *Proceeding of the Sixth International Brick Masonry Conference*, Associazione Nazionale degli Industriali dei Laterizi, Rome, 1982.

[71] Kiamant, R. M. E., *The Chemistry of Building Materials*, Business Books Limited, London, 1970, p. 138.

DISCUSSION

Wm. G. Hime[1] *and Bernard Erlin*[1] *(written discussion)*—Mr. Grimm's paper ignores the extensive world literature on the corrosion of metals in concrete, in spite of the fact that all of it is directly applicable to mortars. That literature comes to but one general conclusion: Corrosion of steel in a portland cement system is triggered either by the presence of substantial amounts of chloride, or by carbonation.

Mr. Grimm ignores the fact that steel and galvanized steel are normally passive when embedded in a portland cement system. For example, he relates corrosion to time of wetness, a factor that deals only with exposed steel. His discussion of contaminants groups together a mixture of chemicals and, for example, refers to hydroxide, calcium, sodium, potassium, and chloride, implying that all of these cause corrosion, whereas steel is perfectly stable in a calcium hydroxide environment unless substantial amounts of chloride are present.

Mr. Grimm's review of the world's literature missed the fact that, as early as 1959, reputable agencies advised against any chloride in mortars that contacted metals. His discussion of his own investigations of several buildings where corrosion occurred generally ignores chloride and carbonation.

Mr. Grimm states that because masonry is more permeable than concrete, greater cover of the steel is required, and reports recommendations of up to 102 mm (4 in.) of cover if the steel is not galvanized. A 102 mm (4-in.) cover cannot be attained in brick masonry construction because the brick portion of the masonry should not be considered as part of it. Hence, the only resolution of the cover problem is to use properly protected metal. Galvanized coatings are recommended in the text; however, in our experience chloride corrodes galvanizing steel. Cadmium coatings are better, and expoxy-coated or stainless steels may be necessary in adverse environments.

From our viewpoint, the masonry industry has failed to heed the mistakes of the bridge engineer and the lessons of the bridge investigators. The latter learned that bridge concrete should not be designed solely on the basis of strength, but on the basis of "quality"—low water-cement ratio to restrict the penetration of chloride. Masonry walls dependent upon embedded metals should likewise be made with a relatively impermeable "chloride-free" mortar. There are many masonry structures in the country that have survived admirably for many decades, and there is no reason why we cannot build all future ones satisfactorily.

C. T. Grimm (author's closure)—Carbonation of mortar increases water per-

[1]Erlin, Hime, Associates Division, Wiss, Janney, Elstner Associates, Inc., Northbrook, IL.

meance and therefore increases corrosion of steel in mortar. The presence of excessive free chloride ions accelerates corrosion of steel in mortar. The paper makes that fact clear. However, chloride and carbonation are not the only corrosive influences. All of the buildings listed in Table 2 are built of conventional mortar and none contained vinylidene chloride latex.

The paper is not confined to a discussion of steel embedded in mortar but includes ties in cavities and shelf angles, which certainly are affected by time of wetness.

The Erlin and Hine statement that a 102-mm (4-in.) cover cannot be attained in brick masonry is in error. Evidence to the contrary abounds in the copious world literature on reinforced brick masonry.

Harry A. Harris[1]

A Method to Determine Efflorescence and Water Permeance of Masonry Mortars

REFERENCE: Harris, H. A., **"A Method to Determine Efflorescence and Water Permeance of Masonry Mortars,"** *Masonry: Research, Application, and Problems, ASTM STP 871*, J. C. Grogan and J. T. Conway, Eds., American Society for Testing and Materials, Philadelphia, 1985, pp. 88–100.

ABSTRACT: A laboratory test procedure to measure the efflorescence and water permeance of masonry mortars is presented. Quantitative and qualitative values of the efflorescence growth compounds can be measured using a technique that accelerates their development in a controlled laboratory environment. Using this same simple test apparatus, a minor change in testing procedures will also provide quantitative data on the water permeance of masonry mortars.

KEY WORDS: efflorescence, masonry, mortar, portland cement, permeability

Two of the most prevalent technical service problems related to masonry mortars are efflorescence and water permeance. These problems have been studied for many years; however, suitable testing procedures had never been developed to measure these qualities. The testing procedure described herein will provide meaningful data on both properties with one simple testing procedure.

The objective used in the design of this test procedure was to provide a laboratory test to measure the masonry qualities affecting these two important problems. It was necessary to provide a testing procedure to simulate the field problems that had been encountered. It should, however, reflect only the conditions related to the materials in question, in this case the mortars being used. To do this, the workmanship in the field and the masonry unit properties should be eliminated, with the test concerned only for the masonry mortar being used. The test results achieved should reflect the differences in properties among the mortars being used.

[1]Director of research, Ash Grove Cement Company, Kansas City, KS 66103.

This laboratory test procedure was designed to provide a simple and cost-effective method of obtaining the information required. A minimum of laboratory technicians' time is required and the cost of equipment and materials are very low. As a part of this cost-effective aspect, the time requirements to complete the test are short and meaningful data reflecting the basic field problem appears to be achieved within 14 days. The test results obtained are reproducible and provide data of meaningful value to the parameters desired. An added benefit to this test procedure may be its adaptation to provide data on other materials in addition to its primary function for masonry mortars.

Test Procedure

The principle used in this testing procedure is to mold a small masonry mortar specimen within a plastic cylinder and allow the mortar to hydrate under standard curing conditions. The specimen is then sealed in a watertight jacket with one end exposed to normal laboratory atmosphere and the other subjected to a head of 25.4 cm (10 in.) of water pressure. The water is allowed to permeate the test specimen; upon reaching the exposed surface, it evaporates and leaves behind an alkali residue identical to efflorescence development under normal field conditions. By supplying a meaningful static head to the mortar specimen, a measure of water permeance can be obtained at the same time the efflorescence is being developed on its exposed surface. After this development has been achieved, the specimen is removed, given a light wash with dilute acid, and the degree of efflorescence measured.

The data obtained from this system is a measure of the water passing through the test specimen. Water permeance has been defined here as water flow per unit area per time expressed as millilitres per square centimetre per day. The efflorescence data will provide specific values for each of the alkali compounds present. It is possible to measure any soluble element present in this system as it is deposited on the surface of the mortar. Since efflorescence is related to a problem of discoloration, the values used herein have been combined as a total quantity of the sodium and potassium present. They are not expressed as an alkali equivalent as is done with most cementitious data, but are expressed as total alkali in milligrams per square centimetre as the final measurement of the efflorescence properties.

The actual construction of the test apparatus for water permeance and efflorescence development uses simple polyvinyl chloride (PVC) plastic pipe fittings. The specimens are cast in PVC pipe with an internal diameter of 3.81 cm (1.5 in.), external diameter of 4.06 cm (1.6 in.), and cut to 7.62 cm (3 in.) lengths. After curing, the test specimen is mounted in a PVC pipe compression fitting capped on one end and equipped with a tubing connection to supply the 25.4 cm (10 in.) water head. A manometer is attached to the system using the simple water column to supply and measure the static head of water. The test specimen and holder are shown in Fig. 1 and the assembly of this completed apparatus is shown in Fig. 2.

The preparation of the test specimens uses a standard ASTM Specification for Mortar for Unit Masonry (C 270) mix design with 3 volume parts masonry sand to 1 volume part masonry cement under standard flow rate (100 to 115%) conditions. The masonry sand used in this procedure should be free of water soluble alkalies and well graded to meet the standard aggregate specifications.

To prepare the specimen, the 7.62 cm (3 in.) PVC pipe sections are mounted to steel or glass plates with sealing wax. The specimen is then molded in a similar fashion as standard 50 by 50 mm (2 by 2 in.) cubes with the exception of using a 9.53 mm (3/8 in.) round steel tamping rod to replace the standard rubber tamper. The top surface of the specimen is then cut off squarely with a steel straightedge and the specimen is cured in a moist cabinet.

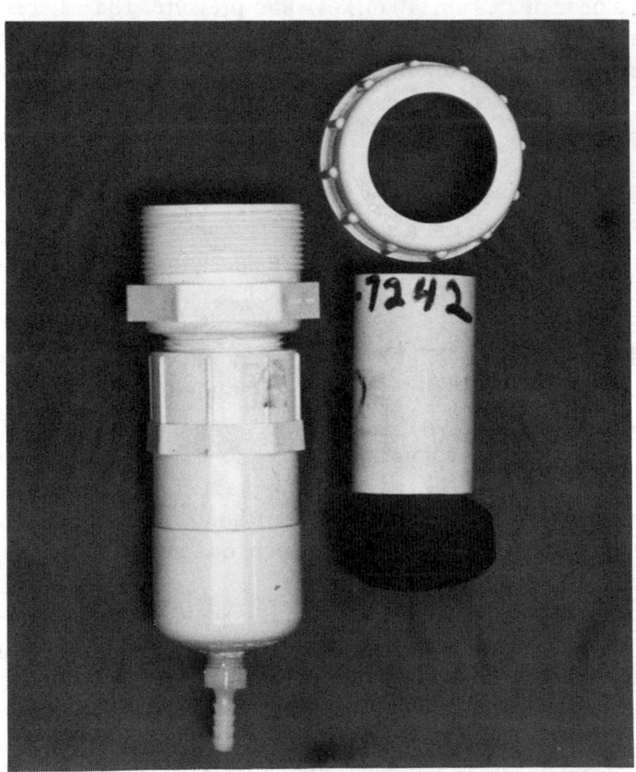

FIG. 1—*PVC pipe test apparatus and specimen.*

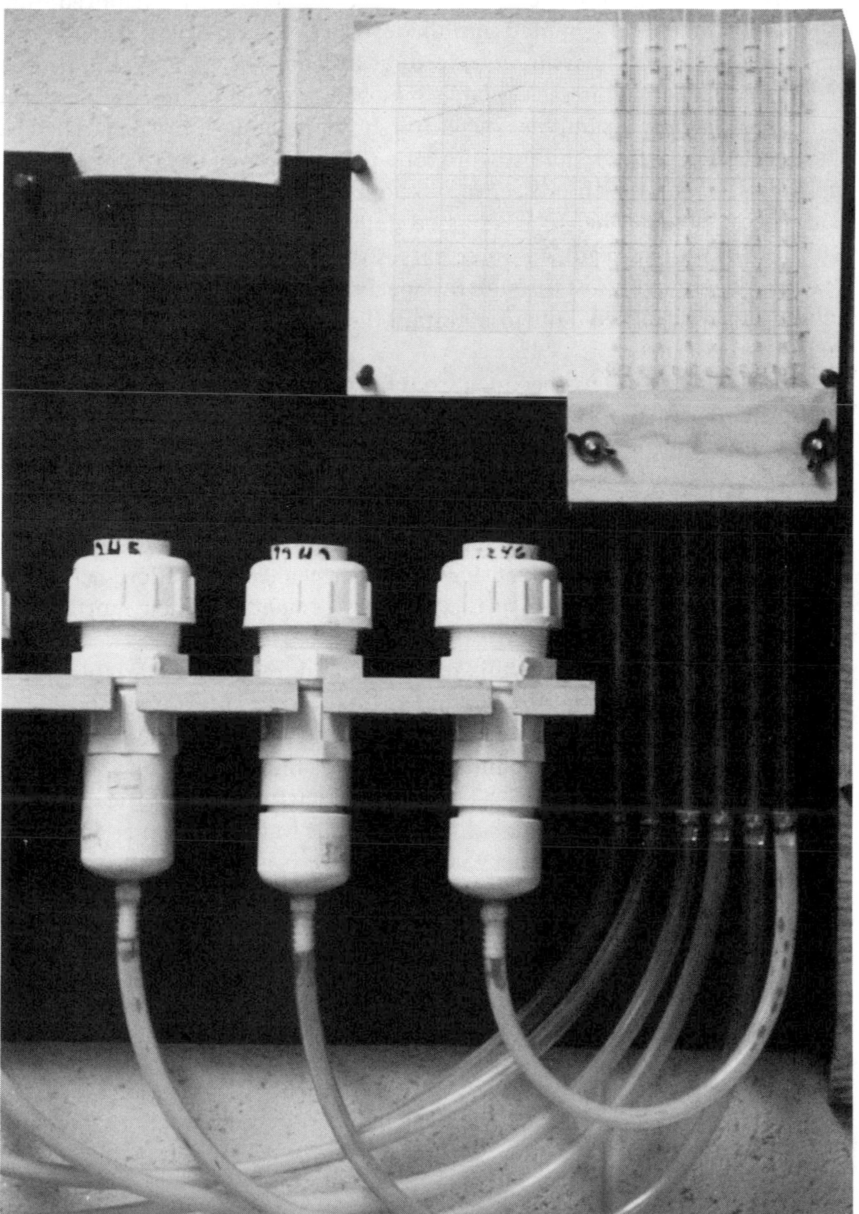

FIG. 2—*Test rack assembly.*

Results and Discussion

At the end of a seven-day curing period, the test specimen is removed from the moist cabinet and mounted into the test rack. Care must be taken in this mounting procedure to assure a watertight fit using the compression fitting. The entrapped air within the system is removed by inverting the specimen and gently shaking any air bubbles out through the manometer system. The test specimen is left under water pressure for a period of seven days before removing it and testing for efflorescence development. During this seven-day period, the manometer height is adjusted daily to maintain a 25.4 ± 1.25 cm (10 ± ½ in.) water head. As water passes through this specimen, the water head will be gradually lost through the specimen, and the manometer must be adjusted or replenished daily to maintain the water head as close as possible to 25.4 cm (10 in.).

After removing the test specimen from the test rack, it is inverted in an acid solution and washed for removal of the alkali content. The details of this efflorescence removing procedure will be found in the appendix of this report.

Typical alkali efflorescence development ranges from 0 to 3 mg/cm^2 of surface area. Water permeance on these same samples ranges from 0.08 to 0.6 mL/cm^2 per day. A few typical masonry cement mortars have been tested in this system; the values obtained for water permeance and efflorescence are shown in Table 1. Properties of the masonry cement and their mortars are shown in Tables 2 and 3.

During the development of this testing procedure, several variables were studied to determine their effect upon the final results. Mortar compositions were varied including the cement: sand ratios and the types of sand and sand gradations used. The use of a standard masonry sand at 3:1 ratio is suggested to comply with mortars under ASTM C 270 and actual field mortars. Curing time on specimens prior to their submission to the water permeance and efflorescence stage has been arbitrarily set for seven days under all test conditions. The length of the test for water permeance and efflorescence development has

TABLE 1—*Water permeance and efflorescence data (development time = 21 days).*

Mortar Sample	Specimen	Water Permeance Flow Rate, mL/cm^2	Efflorescence, Total Alkali, mg/cm^2
G	1	0.16	0.06
	2	0.16	0.00
H	1	0.38	2.31
	2	0.34	1.78
I	1	0.44	1.45
	2	0.26	0.17
J	1	0.59	1.80
	2	0.58	1.84

TABLE 2—*Mortar properties.*[a]

Parameters	Mortar Type					
	A Masonry Cement	*B* Masonry Cement	*C* Masonry Cement	*D* Masonry Cement	*E* Portland-Lime	*F* Masonry Cement
Air content, %	20.4	17.9	18.5	16.0	1.9[b]	18.3
Compressive strength						
7 days, psi	1630	2070	1540	1440	NA	1350
28 days, psi	1960	2500	1740	1970	NA	NA
Fineness						
325 mesh, weight % passing	92.8	93.2	99.1	96.0	NA	89.0
Total alkali content						
Na_2O, weight %	0.17	0.18	0.08	0.18	NA	NA
K_2O, weight %	0.30	0.50	0.19	0.27	NA	NA
Water soluble alkalies						
Na_2O, weight %	0.023	0.033	0.026	0.040	0.051	0.110
K_2O, weight %	0.060	0.140	0.130	0.038	0.111	0.460
Water retention, *T*	81.5	81.9	80.4	60.6	88.6	80.0

[a]Data were obtained using standard ASTM test methods.

[b]All air contents were run by ASTM Specification for Masonry Cement (C 91-78) except Mortar *E* which was tested using commercial sand meeting ASTM Specification for Aggregate for Masonry Mortar (C 144-81). Compressive strengths and water retention were tested with sand prescribed by ASTM C 91-78.

TABLE 3—*Mortar properties.*[a]

Parameters	Mortar Type						
	G Masonry Cement	*H* Masonry Cement	*I* Masonry Cement	*J* Masonry Cement	*K* Masonry Cement	*M* Portland-Type I	*N* Portland Latex Modified
Air content, %	19.1	21.4	21.0	17.3	18.2	8.6	10.0
Compressive strength							
7 days, psi	1790	920	1620	1250	1650	4890	4720
28 days, psi	NA	1220	1940	1700	2010	6150	5350
Fineness							
325 mesh, weight % passing	98.8	83.8	92.5	96.5	95.8	95.9	NA
Total alkali content							
Na_2O, weight %	NA	0.20	0.13	0.12	0.12	0.31	NA
K_2O, weight %	NA	0.44	0.65	0.24	0.43	0.27	NA
Water soluble alkalies							
Na_2O, weight %	0.10	0.027	0.045	0.011	0.037	NA	NA
K_2O, weight %	0.38	0.166	0.426	0.020	0.258	NA	NA
Water retention	78.0	83.8	84.5	78.0	74.8	NA	NA

[a]Data were obtained using standard ASTM test methods with air contents, compressive strengths and water retention tested with sand prescribed by ASTM Specification for Masonry Cement (C 91-78), except mortars *M* and *N*, which were tested with commercial sand meeting ASTM Specification for Aggregate for Masonry Mortar (C 144-81).

been varied using 7, 21, and 28 days. Data on 11 different masonry cements and their mortars is presented in the attached appendix.

Through periodic measurements of the water permeating through the test specimen, a plot of this usage versus time has been provided in Figs. 3 through 6. Figure 3 provides typical data on four different mortar materials. The masonry cement mortar and portland-lime mortar are nearly equal with just over 30 mL of water passing through the specimens in 28 days. The water passage of regular Type I portland cement mortar for this same period, however, is less than 10 mL. A specialized portland-based mortar containing la-

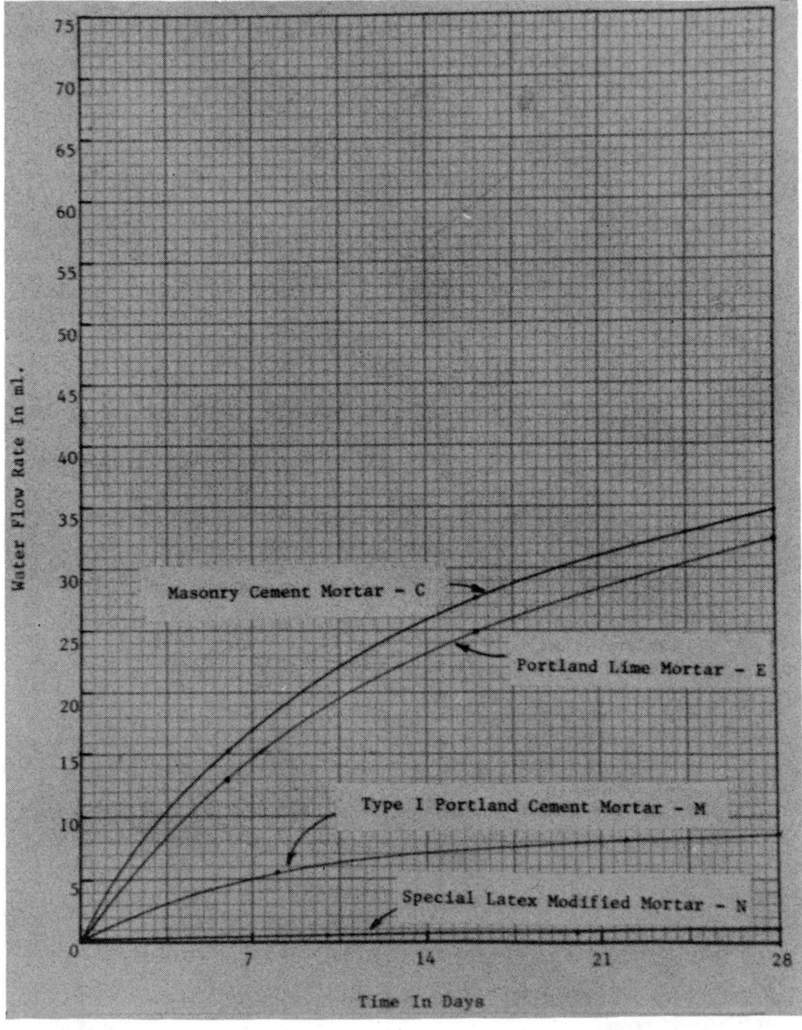

FIG. 3—*Water permeance of various mortar compositions.*

tex has shown very little water permeance, with only 1 mL passing through the system in the 28-day period. Figure 4 shows duplicate tests on two masonry cement mortars with good correlation between both duplicate specimens tested. Figure 5 represents typical data on one masonry cement tested in duplicate for a 7-day period and also for a 28-day period with all results showing only a minor variation. In Fig. 6, two masonry cement mortars were tested in duplicate, one again showing excellent duplication while the second specimen provided a wide variability in its water permeance results. This lack of duplication indicated in Fig. 6 on the masonry cement *I* has been experienced on

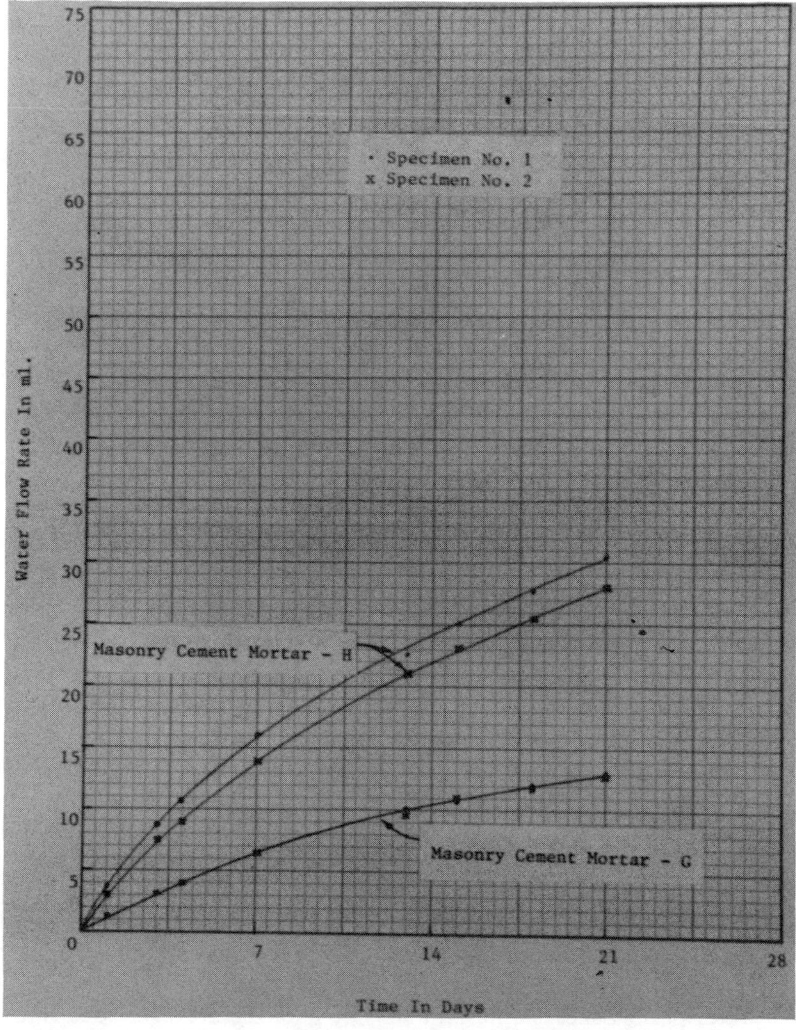

FIG. 4—*Water permeance of masonry cement mortars.*

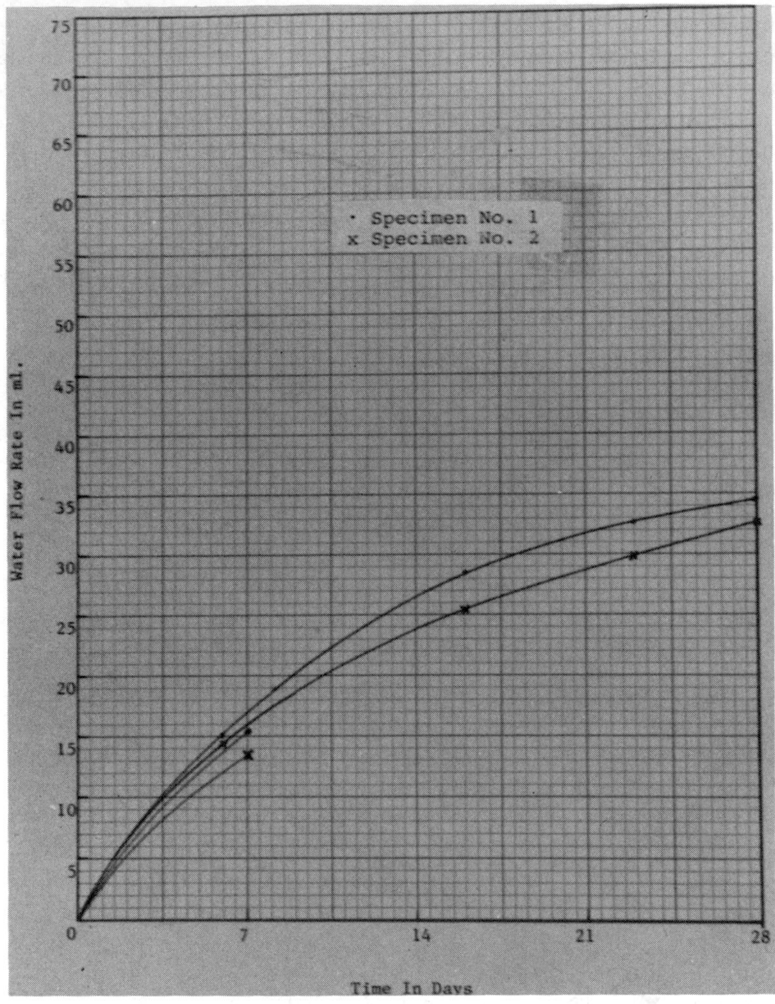

FIG. 5—*Water permeance of masonry cement mortar*—B *at 7 and 28 days.*

other specimens and is assumed to be due to improper compaction of the mortar during the preparation of a test specimen. Examination of the test specimens showing these excessive variations have shown void areas within the mortar but no evidence of leakage between the specimen and the PVC pipe.

In reviewing results of the water permeance as indicated in the foregoing figures, it will be noted that the flow rate through the test specimen decreases with time. This flow rate reduction, as indicated by an average of the 28-day rate of water permeance, is approximately 50% of the average 7-day rate of water permeance. This decrease in flow rate is typical in all specimens and is

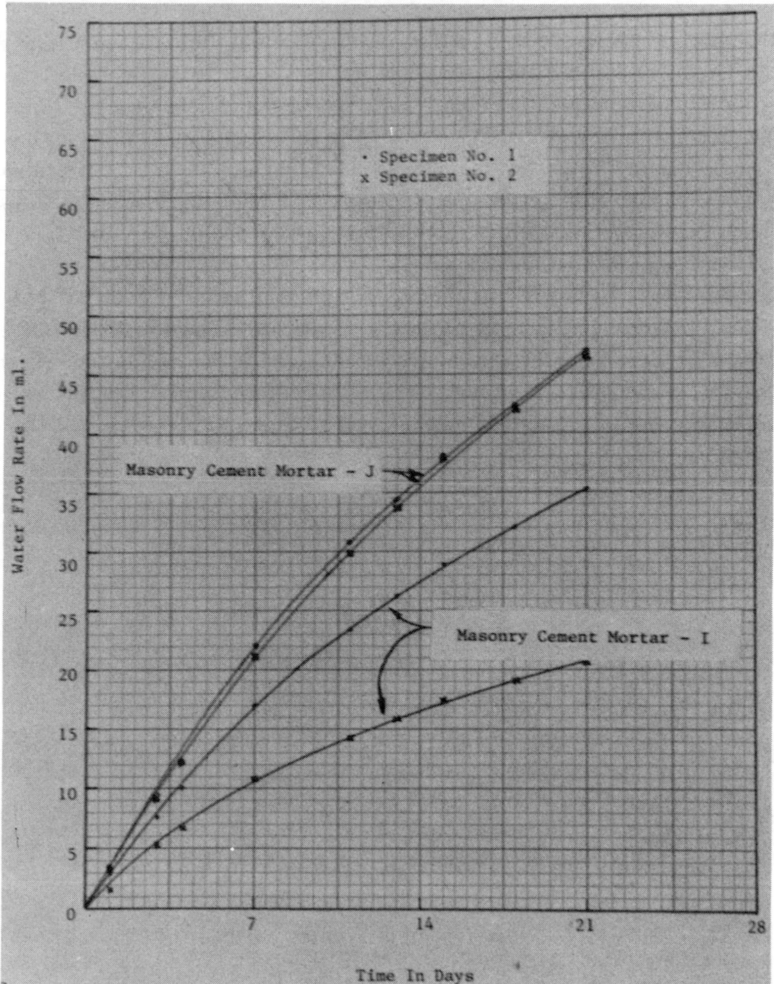

FIG. 6—*Water permeance of masonry cement mortars.*

assumed to be due to the gradual closing of the capillaries within the structure as the cementitious material hydrates and thus reduces its permeance to water passage. A brief review of the data thus far obtained has shown no apparent correlation between water permeance and other mortar characteristics such as air content or water retention.

Efflorescence measurements, based on alkali content on the surface of the specimens, have shown an increase with time spent in the water permeance apparatus. Reviewing the 7- and 28-day data, an average 28-day development is approximately 38% greater than that of the 7-day development. No apparent correlation could be found between the efflorescence data thus far developed and other characteristics of the masonry cement mortars, including that of total or water-soluble alkali content.

Water permeance measurements and efflorescence developments indicate some correlation; however, unexplained variations exist with mortars having both high and low water permeance. Mortar *G* has a low water permeance of 0.16 mL/cm^2/day and little, if any, efflorescence, while mortar *C* shows a similar water permeance but a moderate efflorescence of 0.78 and 1.37 mg/cm^2 at 7 and 28 days respectively.

Based on the tests run during this development, it is suggested that a seven-day curing period be followed by a seven-day period for water permeance and efflorescence development. Excessive variations are shown in the efflorescence data on sample *I* of Table 5 between specimens *1* and *2*, and on samples

TABLE 4—*Water permeance and efflorescence data.*[a]

Parameters	Mortar							
		B Specification						
	A	1	2	*C*	*D*	*E*	*M*	*N*
Water/cementitious ratio	0.55	0.54	0.54	0.53	0.52	0.74	0.47	NA
Water permeance								
total flow 7 days, mL	15.0	15.4	13.3	18.0	15.1	10.7	NA	NA
total flow 28 days, mL	42.9	34.5	32.6	34.7	25.8	32.0	9.0	1.0
flow rate 7 days, mL/cm^2/day	0.19	0.19	0.17	0.22	0.19	0.13	NA	NA
flow rate 28 days, mL/cm^2/day	0.13	0.11	0.10	0.11	0.08	0.10	0.028	0.003
Efflorescence								
Alkali development 7 days								
Na_2O, mg	6.1	2.4	2.3	3.05	3.6	5.6	NA	NA
K_2O, mg	5.6	4.2	3.8	7.8	4.0	2.0	NA	NA
Alkali development 28 days								
Na_2O, mg	6.5	2.0	3.2	6.0	4.0	7.3	NA	NA
K_2O, mg	8.0	3.0	6.4	11.5	5.6	2.9	NA	NA
Control—28 days								
Na_2O, mg	0.8	0.5	0.45	0.6	0.5	0.5	NA	NA
K_2O, mg	1.0	0.5	1.0	1.3	0.55	0.55	NA	NA
Total alkali 7 days, mg/cm^2	0.87	0.49	0.41	0.78	0.57	0.57	NA	NA
Total alkali 28 days, mg/cm^2	1.11	0.35	0.71	1.37	0.75	0.80	NA	NA

[a]Mortars for water permeance and efflorescence tests were prepared using ASTM Specification for Mortar for Unit Masonry (C 270-82) and commercial sand meeting ASTM Specification for Aggregate for Masonry Mortar (C 144-81). Total alkali is shown as $Na_2O + K_2O$ and not as equivalent Na_2O.

B shown in Table 4 where 7-day values exceed 28-day values. To provide reproducible data to reflect these properties, three specimens and one control are suggested for each mortar on future work.

The control samples are prepared for the efflorescence tests by curing a regular specimen in the moist cabinet for the full curing and efflorescence development period. The control is then subjected to the same efflorescence removal procedures as the test specimen. The minor problems noted on variations between duplicate test specimens should be corrected by the addition of a third specimen. Although test times were extended for several specimens in this development series, the data obtained at seven days appears to have always reflected the results that were later found with the test extension.

Conclusion

This test apparatus is inexpensive and simple to construct and assemble. The data obtained will reflect the properties of the masonry mortar in question, and will not be hampered by other factors such as the masonry units or construction techniques not directly related to the masonry cements or mortars involved. Major differences have been shown between mortars containing different masonry cements presently available in the marketplace, and these

TABLE 5—*Water permeance and efflorescence data.*[a]

	Mortar					
Parameters	*F*	*G*	*H*	*I*	*J*	*K*
Water/cementitious ratio	0.55	0.52	0.52	0.52	0.52	0.52
Water permeance						
total flow—21 days						
Specimen 1, mL	49.2	12.7	30.7	35.1	46.9	41.3
Specimen 2, mL	34.8	13.1	27.0	20.8	46.4	28.9
Flow rate—21 days						
Specimen 1, mL/cm^2/day	0.61	0.16	0.38	0.44	0.59	0.52
Specimen 2, mL/cm^2/day	0.44	0.16	0.34	0.26	0.58	0.36
Efflorescence						
alkali development—21 days						
Specimen 1—Na_2O, mg	20.3	1.0	13.15	2.6	6.9	8.55
K_2O, mg	19.15	2.35	14.5	15.15	14.7	11.35
Specimen 2—Na_2O, mg	15.35	0.45	7.9	0.6	7.15	6.35
K_2O, mg	19.3	1.7	13.65	2.45	14.95	7.65
control—Na_2O, mg	0.45	0.6	0.35	0.25	0.2	0.2
K_2O, mg	1.0	2.0	0.75	0.8	0.7	0.45
Total alkali						
Specimen 1, mg/cm^2	3.31	0.06	2.31	1.45	1.80	1.67
Specimen 2, mg/cm^2	2.89	0.00	1.78	0.17	1.84	1.16

[a]Mortars for water permeance and efflorescence tests were prepared using ASTM Specification for Mortar for Unit Masonry (C 270-82) and commercial sand meeting ASTM Specification for Aggregate for Masonry Mortar (C 144-81). Total alkali is shown as $Na_2O + K_2O$ and not as equivalent Na_2O.

differences indicate the potential for improvements of masonry cements through the use of the test procedure. The results between masonry cement mortars *G* and *H* are a good example showing more than a 3.6:1 difference in water permeance and a 60:1 difference in efflorescence development. Tests on the portland-lime mortar, portland Type I mortar, and special latex modified mortar shows the use of this test method for numerous other materials and mortar combinations.

APPENDIX

Dilute HCl Extraction of Efflorescence Alkalies from Mortar Specimen

1. Place 100-mL dilute acid (10% HCl) into 250-mL beaker; add a Teflon magnet and stir solution very slowly with a magnetic stirrer, without creating a vortex.
2. Immerse concrete surface to be tested just below the acid surface (approximately 1 cm) and time for 1 min.
3. After 1 min, remove sample and rinse twice with 10-mL portions of deionized water.
4. Filter using fast paper (541) into 500-mL volumetric flasks. Rinse paper six times and dilute to the mark. Run a blank with all samples.
5. Prepare 1.0, 5.0, 10.0, 25.0, and 50.0 ppm Na_2O and K_2O standard solutions in same containers.
6. Run the filtered solution on the AA by establishing a linearity of standards and then bracketing the unknowns. Subtract blank parts per million directly from unknown parts per million prior to calculation. Since no sample weight is involved, report total milligrams K_2O and Na_2O according to the following equation:

$$\frac{\text{ppm in 500 mL}}{2} = \text{ppm in 1000 mL} = \text{total in mg}$$

Frank L. Yi[1] *and Ramon L. Carrasquillo*[1]

A Study of the Thermal Behavior of Brick Under Service Conditions in a Structure

REFERENCE: Yi, F. L., and Carrasquillo, R. L., **"A Study of the Thermal Behavior of Brick Under Service Conditions in a Structure,"** *Masonry: Research, Application, and Problems, ASTM STP 871*, J. C. Grogan and J. T. Conway, Eds., American Society for Testing and Materials, Philadelphia, 1985, pp. 101–119.

ABSTRACT: The main purpose of this paper is to present results of a three year research project conducted at The University of Texas at Austin on the development of a test procedure for determining the coefficient of thermal expansion of brick under service conditions in a structure. Many currently used masonry design and construction practices are the result of past experience. With the development of new design methods and construction techniques for brick masonry, a better understanding of the thermal behavior of brick is needed. Unaccounted thermal-induced movements in brick masonry have been the reason for observed cracking in brick masonry walls in many structures in service today.

In the study reported herein, the effects of moisture conditions including dry, 50 percent of saturation, and saturated, and of number of temperature test cycles ranging from −18 to 60°C (0 to 140°F) under a given moisture condition on the observed thermal behavior of new (unused) bricks in three directions (length, height, and width) was determined. Strain gages bonded to the surface of each specimen in the direction of measurement were used to monitor temperature-induced movements in the brick continuously during testing. The effect of location of the strain gages on the surface of the brick in each direction on the observed behavior was also determined. The thermal behavior of several brick specimens removed from existing structures after years in service was also observed in order to correlate the results obtained using new brick specimens.

A proposed test method in the form of ASTM standards was developed for determining the coefficient of thermal expansion of bricks under expected service conditions in the structure. Information obtained from this test is needed by practicing engineers in the design of expansion joints in brick masonry to ensure safe and durable structures.

KEY WORDS: brick, coefficient of thermal expansion, thermal behavior, moisture conditions, temperature cycles

[1] Associate Professor of Civil Engineering, and Professor of Civil Engineering, respectively, The University of Texas at Austin, Austin, TX 78758.

Brick is one of the oldest building materials. Consequently, many common practices in brick masonry design and construction are the results of many years of experience rather than of theoretical analysis. Thermal expansion data for brick have important uses both in design and in the development of ceramic theory. In the building and construction industry, such data are used in the development of a structural unit which will withstand repeated temperature changes. From the material standpoint, the thermal expansion behavior of the brick is one of its fundamental properties.

The change in length of building bricks due to temperature change is one of the factors often blamed for cracking of brick masonry walls. The coefficient of thermal expansion of brick is the single most important parameter needed for predicting the thermal behavior of brick masonry.

During the past decade, extensive research has been conducted on brick masonry but not much work has been done on methods and means of determining the coefficient of thermal expansion of brick [*1-6*]. Many unknowns need to be understood before arriving at a proposed test method for determining the coefficient of thermal expansion of brick. The work reported herein deals with the understanding of such factors as moisture condition, temperature range, number of temperature cycles, and location of the strain measuring instrumentation on the surface of the brick on the observed thermal behavior of new (unused) brick and of brick removed from existing structures.

Experimental Program

Brick Specimens

Eight extruded stiff-mud clay bricks were used in the study. Four bricks, A through D, refer to new bricks as received from the supplier. The other four bricks, E through H, were old building bricks removed from different existing structures after at least 10 years in service. Relevant physical properties of the bricks tested are given in Table 1.

TABLE 1—*Physical properties of bricks used in this study.*[a]

Brick Specimen	Compressive Strength (ASTM C 67-79), N/mm^2	5-Hr Boil Absorption (ASTM C 67-78), %	24-Hr Cold Absorption (ASTM C 67-78), %	Saturation Coefficient (ASTM C 67-78)
A, B, C, D	93.8	2.19	0.63	0.28
E, F	62.8	2.21	1.68	0.78
G	97.2	6.55	6.31	0.96
H	80.0	3.48	1.42	0.37

[a] All the values shown in the table are the average value of at least ten test results.

Method of Testing

The main objective of this study was to examine the three-dimensional thermal expansion behavior of brick. Electrical resistance strain gages bonded to the surface of the brick were used to continuously measure length change in a given direction in the brick. Testing consisted of subjecting each brick specimen to a predetermined number of temperature cycles ranging from −18 to 60°C (0 to 140°F) at a given moisture content. Since the coefficient of thermal expansion of brick is not always a linear function with temperature and some bricks have coefficients with multiple modes, or peak values of different magnitudes, or both, length change measurements were made approximately every 13°C (20°F) temperature change within the test temperature range in order to determine the variation in coefficient of thermal expansion of brick with temperature.

Length change measurements in a given direction (length, height, or width) were made using two electrical resistance strain gages connected in series and glued to the surface of the brick in that direction. A total of twelve strain gages, four in each direction, were glued to each brick specimen tested. The strain gages were then connected to a switch and balance unit using a half bridge circuit as shown in Fig. 1. The compensating arm of the circuit was formed by a similar gage arrangement glued to a piece of fused quartz having a known coefficient of thermal expansion of 0.45×10^{-6}/°C.

A thermocouple wire (iron-constantan) was attached to the surface of each brick being tested to monitor the temperature of the bricks. Previous tests [*2, 7*] have shown that the difference between the average temperature within a brick and the surface temperature measured as described in this study is negligible. In addition, a similarly instrumented piece of aluminum of a known coefficient of thermal expansion was subjected to the same test conditions as the bricks in order to check the validity of the test procedure and proper functioning of the test equipment. The testing arrangement is shown

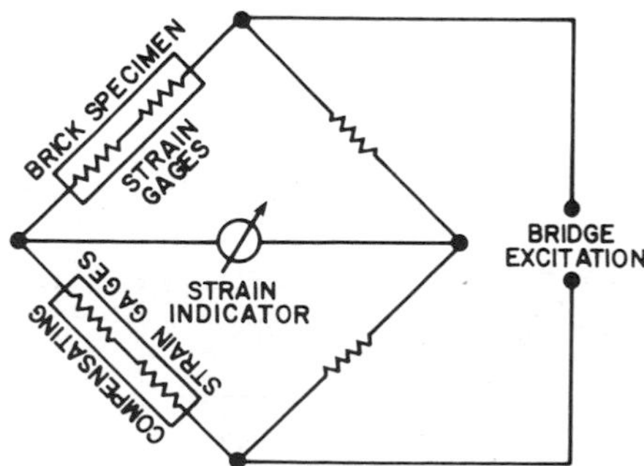

FIG. 1—*Wheatstone bridge circuit connection (half-bridge).*

in Fig. 2. A complete description of the specimen and testing conditions is given in Table 2.

The linear coefficient of thermal expansion was calculated between successive temperature T_1 and T_2 as follows:

$$\alpha = (\epsilon_2 - \epsilon_1)/(T_2 - T_1) + \alpha_c$$

where

α = linear coefficient of thermal expansion,
ϵ_2 = strain reading at temperature T_2,
ϵ_1 = strain reading at temperature T_1,
T_1, T_2 = temperature values, and
α_c = linear coefficient of thermal expansion of compensating specimen (fused quartz).

Experimental Results and Discussion

Strain Gage Location

To determine the effect of the location of a strain gage on the surface of the brick in a given direction (length, width, or height) on the observed thermal behavior of the brick, some bricks were instrumented with more than one set of strain gages in each direction, as shown in Fig. 3. In addition, some of the strain gages on a specimen were not bonded on symmetrically opposite locations on the faces of the brick but were bonded on the corners or edges of the bricks. Nevertheless, the same consistent thermal coefficient magnitude and distribution characteristics were observed at different strain gage locations on the surface of the brick aligned in any given direction X, Y, or Z based on the test results, as shown in Fig. 4. This implies that the thermal behavior of the bricks tested is independent of the location of the gage on the surface of the

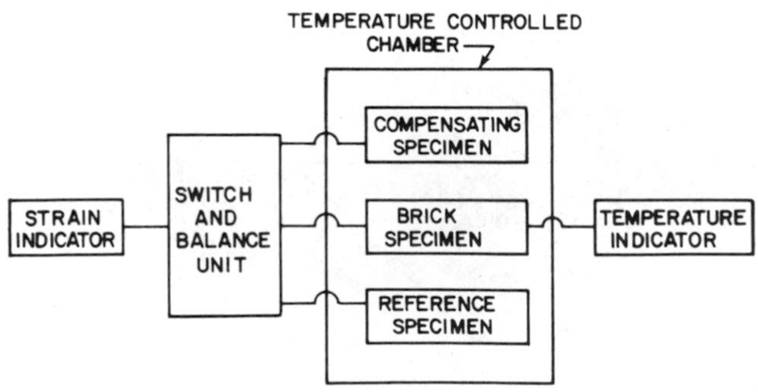

FIG. 2—*Test arrangement.*

TABLE 2—*Specimen description and testing conditions.*

Brick Specimen	Moisture Content	Number of Test Cycles	Measurement Direction
A, B, C, D	dry	3	Three orthogonal directions (length, height, and width).
	50% saturated	3	Two pairs of strain gages in each direction.
	fully saturated[a]	3½	
E, F, G, H	uncontrolled	2	Three orthogonal directions (length, height, and width).
	uncontrolled	2	One pair of strain gages in each direction.
TEMPERATURE RANGE, °F (°C)			
0 (−17) to 20 (−6.7) to 40 (4.5) to 75 (23.9) to 95 (35.0) to 115 (46.1) to 140 (60)			

[a] For the fully saturated condition, testing continued for an additional half cycle after the third complete cycle.

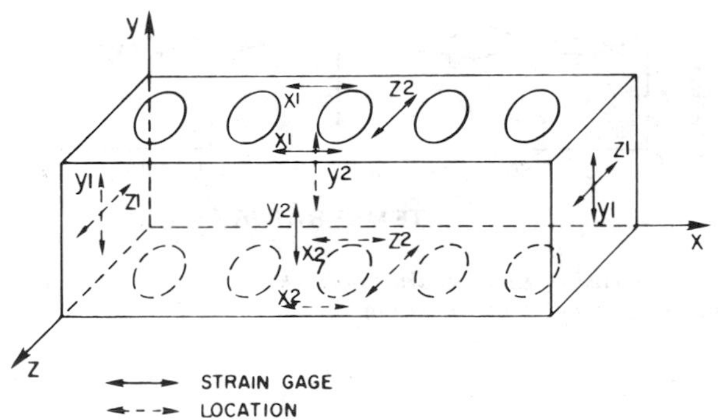

FIG. 3—*Strain gage locations on brick specimens in three directions.*

brick. For this reason, the average thermal coefficient value for the three orthogonal directions is used in plotting the test results for each brick specimen.

Temperature Range

As shown in Fig. 5, the coefficient of thermal expansion of brick is not always a linear function with temperature. For this reason, the coefficient of thermal expansion of a brick should be measured over the temperature range to which the brick will be subjected in service. In addition, it was observed that different specimens of the same type of brick tested under the same conditions, moisture and temperature range, show similar thermal coefficient

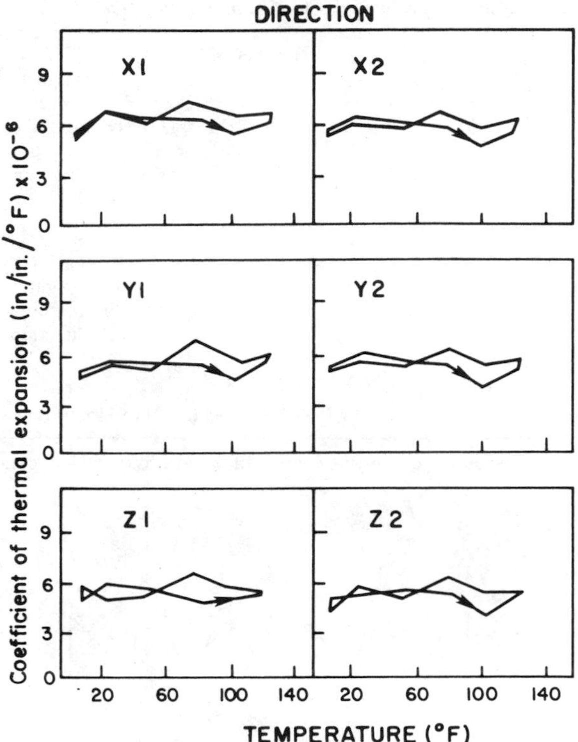

FIG. 4—*Typical observed thermal behavior of bricks in different directions using different strain gage locations X, Y, and Z, shown in Fig. 3.*

distribution behavior but probably with different magnitudes of coefficient of thermal expansion, as shown in Fig. 5 for bricks A, B, C, and D.

Moisture Condition and Number of Test Cycles

New Bricks—Tests were conducted under three different moisture conditions, dry, 50% saturation, and fully saturated, using new (unused) bricks. The average moisture content of these bricks at saturation was 1.8 percent of the dry weight of the brick. Once the brick specimens reached equilibrium at the test moisture content, they were sealed inside plastic bags to prevent any significant change in moisture content of the brick during testing. The specimens were subjected to at least three temperature cycles until a stable thermal behavior was observed. From this study, it was found that the thermal behavior of new bricks was affected significantly by the moisture content of the bricks.

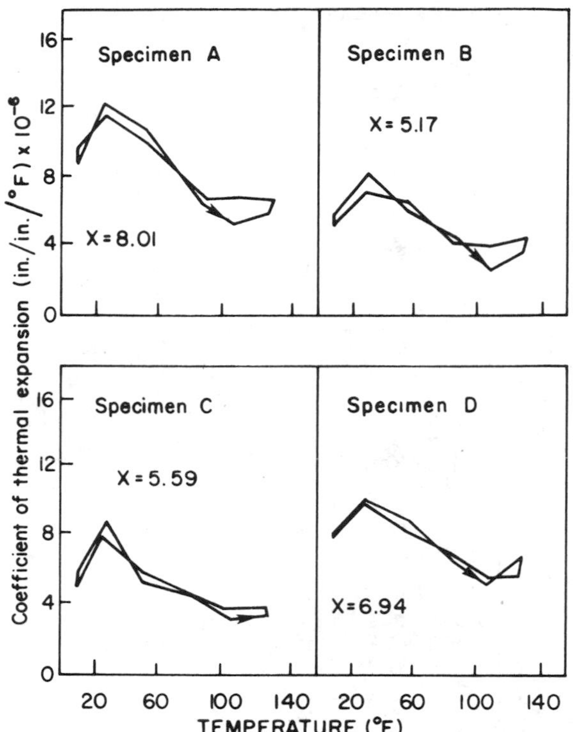

FIG. 5—*Typical observed thermal behavior of the same type of brick specimens under similar test conditions, moisture, and temperature range.*

(*a*) For the dry test condition, since there was no water in the brick throughout the three test cycles, the thermal behavior reached a stable condition from the very first test cycle. This can be seen from the thermal coefficient distribution diagram in Fig. 6. Under this test condition, the number of test cycles has no effect on the observed thermal expansion behavior. No significant difference was observed between the magnitude of the mean thermal expansion coefficient with the number of test cycles in the dry test condition. From these results it can be concluded that only one test cycle is needed for new bricks to reach a stable expansion behavior under dry conditions.

(*b*) For the 50% saturation test condition, a large difference in thermal behavior of the bricks was observed between the first and the second temperature cycles. Higher standard deviation values were found when comparing the mean thermal expansion coefficient values for each test cycle for the 50% saturation test with the standard deviation values of the dry test, especially for the first cycle. This behavior was attributed to the fact that these were new bricks experiencing for the first time the thermal expansion action with 50%

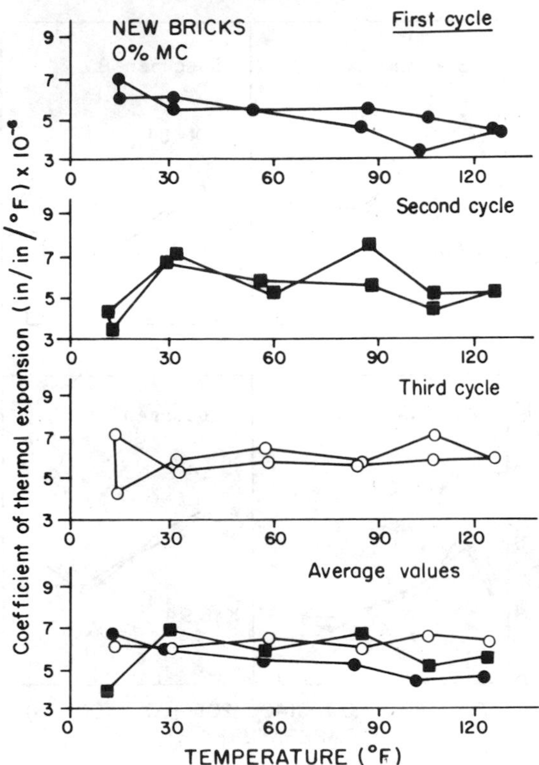

FIG. 6—*Typical mean coefficient of thermal expansion test data for new bricks under dry conditions.*

of the voids filled with water. As a result, the relative thermal-induced movement between the brick, water, and ice will cause the observed variation in coefficient of thermal expansion with temperature. The effect of the water in the brick decreased as the number of test cycles increased. This behavior is shown in Fig. 7. From these results, it was concluded that the presence of water in the brick does have a significant effect on the thermal expansion behavior of the brick. As a result, more than two test cycles are needed for the 50% saturation tests using new bricks to reach a stable thermal expansion behavior. As before, no significant difference was observed between the magnitude of the mean thermal expansion coefficient with the number of test cycles in the 50% saturation test condition.

(*c*) For the fully saturated condition, a large variation in the thermal coefficients was found in the first two temperature test cycles. This can be clearly seen from Fig. 8. Higher standard deviations in the mean coefficient of thermal expansion values were found in the first two test cycles when compared with the values in the dry test. The reason for this was that the voids in the

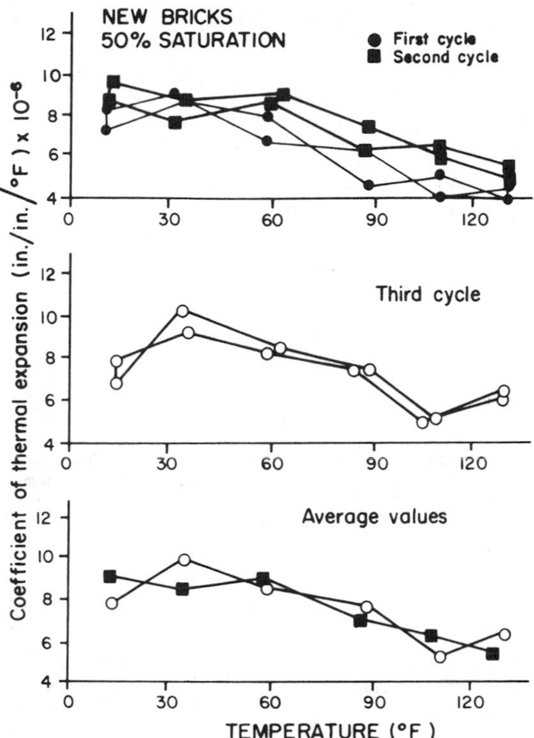

FIG. 7—*Typical mean coefficient of thermal expansion test data for new bricks under 50 percent saturation.*

brick were fully saturated with water and once the brick started to expand or contract, the inner structure of the brick was disrupted by the relative movement between the brick, water, and ice. This resulted in the unstable thermal movement observed in the first several temperature cycles. However, with increasing number of test cycles, a stable thermal behavior was reached. The magnitude of the mean coefficient of thermal expansion measured during test cycles subsequent to achieving a stable thermal behavior in the bucks was constant.

Old Building Bricks

For the old building bricks E, F, G, and H, the moisture content was not controlled throughout the test. All the standard deviation values in the two test cycles for all bricks were smaller than the values in the 50% saturation and fully saturated tests and of about the same magnitude as in the dry test. The behavior shown in Fig. 9 indicates that the number of test cycles has no effect on the thermal expansion behavior of old building bricks.

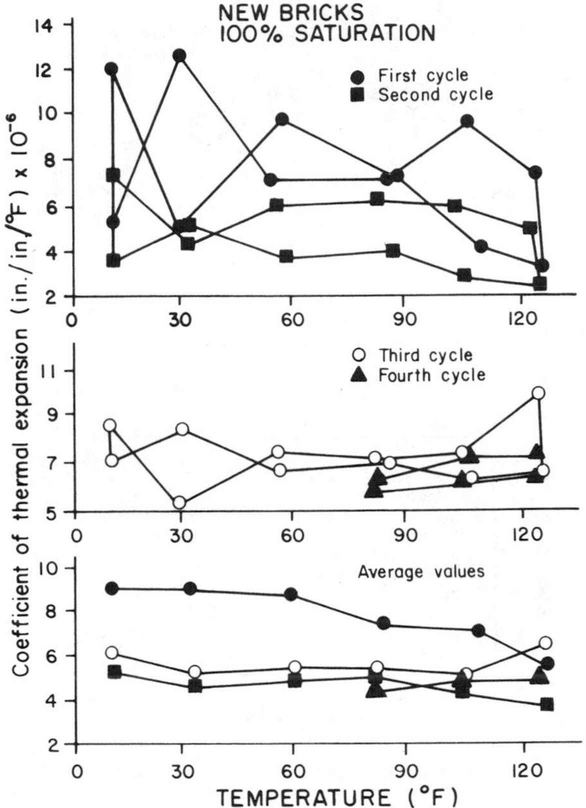

FIG. 8—*Typical mean coefficient of thermal expansion test data for new bricks under saturated conditions.*

It should be noted that these bricks had experienced many cycles of temperature change, and a stable thermal expansion behavior had already been reached. For this case, one test cycle is enough for determining the thermal coefficient of the old building bricks.

Proposed Test Method

Based on the test results presented herein, a proposed test method for determining the coefficient of thermal expansion of brick using electrical resistance strain gages was developed. This test method is currently under review by ASTM Committee C-15 on Manufactured Masonry Units.

Features of the proposed test procedure include the following:

1. The temperature range should consider the weather conditions to which a building brick will be subjected in service. In this study, a temperature range from −17 to 60°C (0°F to 140°F) was used.

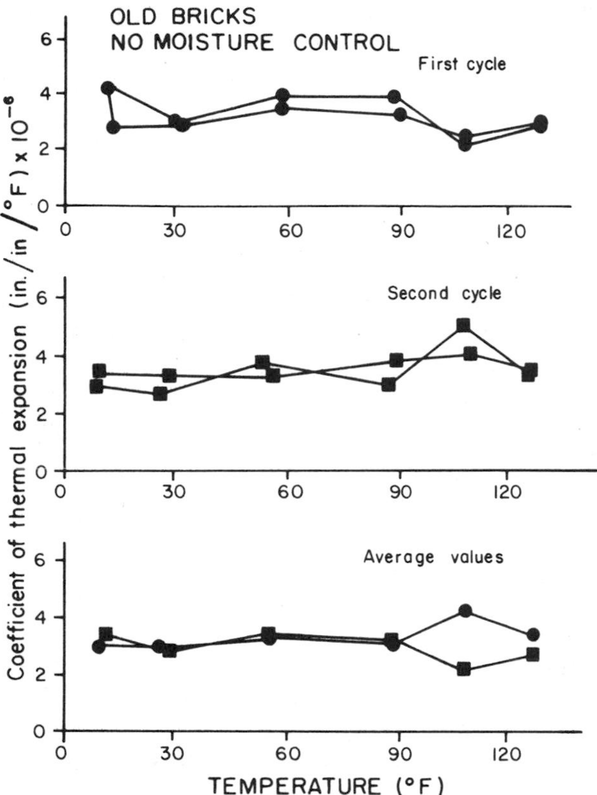

FIG. 9—*Typical mean coefficient of thermal expansion test data for old building bricks tested under no moisture control conditions.*

2. The moisture content of the brick should include conditions from dry to saturated which might occur in different areas or different parts of a structure. Three different moisture contents, dry, 50% saturation and fully saturated, were considered.

3. The number of temperature test cycles should be more than one in order to assess the effect of repeated temperature cycles on the brick behavior, as would be the case of a brick in an actual structure.

4. Since the coefficient of thermal expansion of the brick is not always a linear function with temperature and some bricks have coefficients with multiple modes, or peak values of different magnitudes, or both, the test method should permit a continuous measurement of the thermal coefficient of a brick specimen over small temperature ranges.

5. The coefficients of thermal expansion are frequently very small, which emphasize the need for great precision.

The proposed test method consists of a modification of an existing proposed test method titled "Linear Coefficient of Thermal Expansion of Natural Building Stone with Bonded Strain Gage Extensometers" [*8*]. Some modifications resulting from the study reported herein have been incorporated into the test method to make it applicable to testing brick specimens.

Conclusions

The test procedure presented in this paper provided the necessary information to calculate the coefficient of thermal expansion of brick in three orthogonal directions, length, height, and width. In addition, it provided for the evaluation of the effect of strain gage location, expansion direction, moisture content of the brick, and the number of test cycles on the measured value for the coefficient of thermal expansion. Based on the results from this study, the following conclusions are drawn:

1. The coefficient of thermal expansion of brick is not linear with temperature and has multiple modes or peak values.
2. The thermal expansion behavior of the bricks tested is homogeneous and isotropic with respect to three orthogonal directions, i.e., length, height, and width.
3. The effect of the location of the strain gage on the surface of the brick in a given direction on the measured value for the coefficient of thermal expansion was negligible.
4. Different brick specimens of a given type tested under a given condition show similar thermal coefficient distribution behavior but have different values of the coefficient of thermal expansion for each specimen.
5. No significant effect of moisture content and number of test cycles on the mean value of the coefficient of thermal expansion of bricks was observed.
6. The thermal expansion behavior of new brick was affected significantly by the moisture content condition of the brick.
7. For determining the thermal behavior or the coefficient of thermal expansion of brick, only one test cycle is needed for old building brick specimens and for new brick specimens under dry test conditions. More than two test cycles are needed for new brick specimens under 50% saturation or fully saturated conditions.

References

[*1*] Jafarzadeh, Alireza, "Three Dimensional Thermal Expansion of Bricks," Master Report, The University of Texas at Austin, Dec. 1980.

[*2*] Davison, J. I., and Sereda, P. J., "Measurement of Linear Expansion in Bricks due to Freezing," *National Research Council of Canada, Journal of Testing and Evaluation*, Vol. 6, No. 2, March 1978, pp. 144–147.

[*3*] Palmer, L. A., "Volume Changes in Brick Masonry Materials," *National Bureau of Standards Journal of Research*, Vol. 6, No. 1003, 1931, p. RP321.

[4] Ross, Culbertson W., "Thermal Expansion of Clay Building Bricks," *Journal of Research of the National Bureau of Standards*, Vol. 27, August 1941, pp. 197–216.
[5] Westman, A. E. R., "The Thermal Expansion of Fireclay Bricks," *Engineer Extension Station*, Bulletin No. 181, University of Illinois at Urbana, 1928.
[6] Loubser, P. J., and Bryden, H. G., "Apparatus for Determining the Coefficient of Thermal Expansion of Blocks, Mortars, and Concrete," *Magazine of Concrete Research*, Vol. 24, No. 79, June 1972, pp. 97–100.
[7] Li-Shing Yi, "A Study of the Thermal Expansion Behavior of Brick," Master of Science Thesis, Civil Engineering Department, The University of Texas at Austin, Dec. 1982, 135 pp.
[8] "Proposed Test Method for Linear Coefficient of Thermal Expansion of Natural Building Stone with Bonded Strain Gage Extensometers," *ASTM Standards*, Part 19, April 1975, pp. 469–473.

Robert I. Carr,[1] Richard D. Woods,[1] and John A. Heslip[2]

Guidelines for Temporary Masonry Wall Support Systems

REFERENCE: Carr, R. I., Woods, R. D., and Heslip, J. A., **"Guidelines for Temporary Masonary Wall Support Systems,"** *Masonry: Research, Application, and Problems, ASTM STP 871*, J. C. Grogan and J. T. Conway, Eds., American Society for Testing and Materials, Philadelphia, 1985, pp. 114-125.

ABSTRACT: Masonry walls are vulnerable to high winds until supported by the completed structure. A common method of temporary support consists of inclined braces restrained at the ground with stakes. Adequate information on the capacity of stakes in this system could not be found in the literature, so research was conducted on the horizontal capacity of stakes in soil. Three stakes were studied: 3.8 by 3.8 cm (2 by 2 in.) and 3.8 by 8.9 cm (2 by 4 in.) wood stakes, and No. 6 reinforcing bars. Each stake was tested at two embedments, 30 cm (12 in.) and 45 cm (18 in.), and at three inclinations to the horizontal, 90°, 60°, and 45°. Stakes were tested in sand, silt, and clay at three consistencies. Loads were applied with a hand-cranked screw mechanism with a proving ring to measure load. Displacement and rotation were measured with two dial gages. A total of 139 tests were performed, and load deflection curves were analyzed at ground level displacements of 3.2 mm (1/8 in.) and 6.4 mm (1/4 in.). Equations based on the theory of subgrade reaction for the soils and rigid behavior of the stakes were formulated to match measured data. For most soils, the stake capacity was found to be insufficient to support reasonable lengths of tall masonry walls in a modest wind, 56 km/h (35 mph); consequently, stakes are not recommended as the horizontal support element for tall masonry walls.

KEY WORDS: masonry, wind (meteorology), walls, stakes

When masonry walls are constructed presently, they are subject to collapse due to high winds until they are permanently supported by a completed structure. Construction personnel working on masonry job sites are subject to serious injury in case of collapse of the walls. Presently, the American National Standards Institute (ANSI) provides in ANSI Standard Safety Requirements for Concrete Construction and Masonry Work (A 10.9-170) that masonry walls shall be temporarily shored and braced until the designed lateral strength is reached, to prevent collapse due to wind or other forces. Other

[1]Professor of Civil Engineering, University of Michigan, Ann Arbor, MI 48109.
[2]Executive vice-president, National Concrete Masonry Association, Herndon, VA 22070.

national codes provide similar broad requirements for temporary bracing. There are no definitive guidelines that can be used today by designers and contractors to prevent wall collapse without radically affecting the economics of masonry walls.

Contractors' common methods use stakes in soil to provide lateral support. There is little knowledge of load capacity of such stakes. Contractors use stakes, but they have no basis for their design except experience: experience which, for the industry, shows frequent collapses. Adjustable steel temporary support systems depend on truss action, supported by vertical reactions at temporary footings, a proven technology. However, they are expensive and easily stolen from the site, particularly if delays in completion of the structure also delay permanent support of the walls.

The Michigan masonry industry has worked with the Michigan Department of Labor to develop a state standard for wall bracing. A concensus solution was developed which used common 3.8 by 23.5 cm (2 by 10 in.) scaffold planks supported by a stake in soil. (Cross-section dimensions of wood are nominal, not actual, for English units. 2 in. nominal = 1.5 in. actual, 4 in. = $3^1/_2$ in., 10 in. = $9^1/_4$ in.; metric dimensions are actual.) Figure 1 shows this design, which includes 3.8 by 8.9 cm (2 by 4 in.) bracing [if needed to support the 3.8 by 23.5 cm (2 by 10 in.) shore against buckling under compression]. The horizontal force of the wind is resisted by the wall foundation at the bottom and the brace near the top. The horizontal force carried by the wall can exceed 48 N/m^2 (5 psf) for a 56 km/h (35 mph) wind with gusts. This would produce a horizontal wind force of over 1470 N/m (100 lb/ft) of wall, of which about 1030 N/m (70 lb/ft) must be carried by the stake. This horizontal force is indicated by *H* in Fig. 1.

A search of literature showed a lack of sufficient information from which we could predict the horizontal load capacity of a stake in soil. The University of Michigan Department of Civil Engineering contracted with the Michigan Department of Labor to test the capacity of stakes and develop a simple model which would be useful for engineering design and development of tables for a wall bracing standard.

The University of Michigan worked with the Masonry Institute of Michigan and members of the Advisory Committee to the Construction Safety Standards Commission for the standard. The Advisory Committee selected the bracing system, the stakes which were tested, and the conditions of use, and they oversaw the test procedures to ensure that results met industry needs.

Procedure

The ultimate goal of the experiments was to establish a method to predict adequately the lateral capacity of short stakes. Lateral capacity of the stakes was chosen as the critical capacity because the mechanism of the wall bracing system being considered produces primarily lateral loads on the stakes.

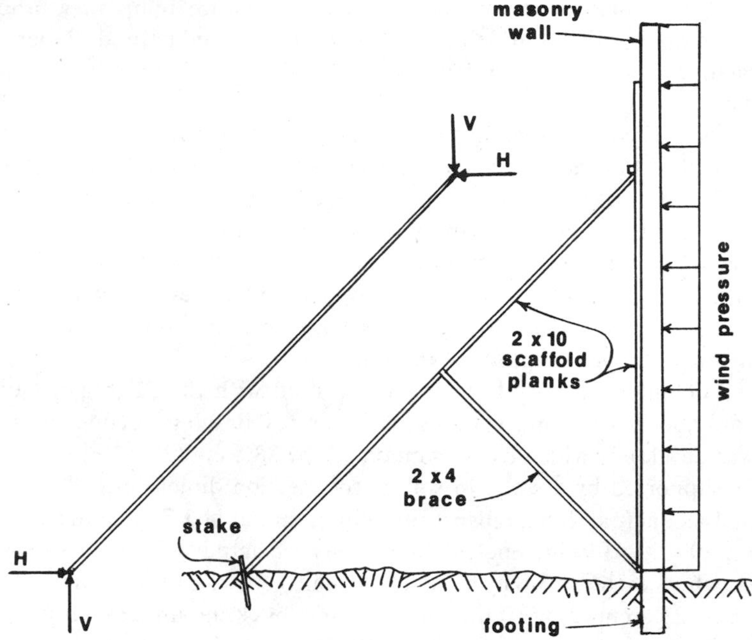

FIG. 1—*Selected temporary brace.*

Three types of stakes were chosen as representative of those most readily available and commonly used in masonry wall bracing systems. They were 3.8 by 8.9 cm (2 by 4 in.) and 3.8 by 3.8 cm (2 by 2 in.) wood stakes and No. 6 (1.9 cm) steel reinforcing bars. The stakes were tested with 30 cm (12 in.) and 45 cm (18 in.) embedment depths. The stakes were driven at angles from the horizontal of 45 deg, 60 deg, and 90 deg and the lateral load was applied at about ground level at angles of 0 deg, 30 deg, and 45 deg to the horizontal.

Three types of "connections" between the stakes and the loading device were tried. One considered of a collar made of 1.3 cm (½ in.) wide steel strap surrounding the stake, Fig. 2, another consisted of a pin directly through the stake, Fig. 3, and the last was a section of 3.8 by 8.9 cm (2 by 4 in.) 25 cm (10 in.) long, pressing directly against the stake, Fig. 4. The first two types were intended to provide an "ideal" connection for analysis, but the results from all three were essentially the same. For this reason most of the tests were performed with the 3.8 by 8.9 by 25 cm board against the stake, which most closely modeled the field condition.

A range of soils was chosen which covered many of the soil types which could be expected at a construction site, both natural occurring and imported. Table 1 lists the soils used and their descriptions and pertinent properties.

FIG. 2—*Collar connection.*

To aid the worker in the field, an attempt was made to develop a simple test which would uniquely identify a soil type and provide an estimation of its strength characteristics. Furthermore, it was intended that the test would properly evaluate the soil throughout the depth of influence of the stake. A penetration test was tested which at first consisted of only one size reinforcing bar pushed into the soil by the weight of a "nominal man," about 760 N (170 lb). The depth of penetration was to be the criterion. As the variety of soils tested reached the total of five, we had tested three rebar penetrometer sizes. We could not make an identification of the soils with a single re-bar size, as shown by the results in Table 2. Because of the conclusions being formulated from the results of these stake tests, that is, generally insufficient capacity for wall bracing, no further refinements were attempted on the soil identification test, and no penetration criteria are being suggested.

Load was applied to the stakes through the hand-cranked screw mechanism shown in Fig. 5. The angle of the screw platform could be adjusted by the location and height of the vertical structure. Horizontal support of up to

FIG. 3—*Pin connection.*

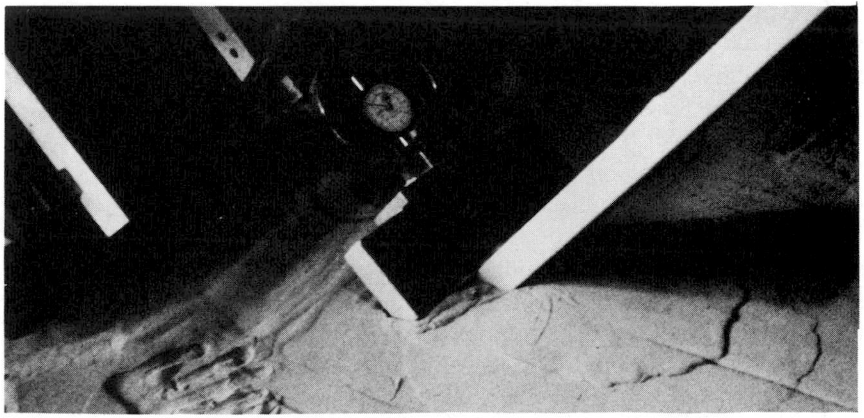

FIG. 4—*3.8 by 8.9 by 25 cm (2 by 4 by 10 in.) connection.*

TABLE 1—*Soil properties.*

Description and Class	Dry Unit Weight, kN/m^3 (lb/ft^3)	Water Content, %	Void Ratio	Angle of Internal Friction ϕ, deg	Unconfined Compressive Strength q_u, kN/m^2 (psf)
Loose to medium dense, fine to medium grain, clean brown sand (SP)	16.4-16.8 (104-106)	2.5-4	0.56-0.58	34-36	. . .
Nonplastic, brown inorganic silt (ML)	15.0-15.3 (95-97)	8-10	0.72-0.80	33-35	. . .
Sandy, silty, grey clay with some pebbles (CL) (natural at two consistencies)	16.1-16.6 102-105	12-18	. . .	. . .	50-120 (950-2500)
Silty, light brown clay (CL) (artificial)	15.8 (100)	24	. . .	. . .	40 (900)

TABLE 2—*Penetration depths using 760 N (170 lb) force.*

	Soil Type				
Bar Size	ML	SP	CL—Stiff	CL—Medium	CL—Soft
1.3 cm (No. 4)	66 cm (26 in.)	33 cm (13 in.)	3 cm (1 in.)	no test	>66 cm (>26 in.)
1.9 cm (No. 6)	25 cm (10 in.)	25 cm (10 in.)	0	20 cm (8 in.)	>66 cm (>26 in.)
2.5 cm (No. 8)	20 cm (8 in.)	23 cm (9 in.)	no test	no test	no test

4500 N (1000 lb.) was supplied either by dead weights (steel and concrete) laid on the base frame or by a screw plate anchor embedded about 1 m (3 ft) deep in the soil and attached to the base frame.

Load on the stakes was measured by a proving ring with a maximum capacity of 4500 N (1000 lb) (see Figs. 2 to 5). Displacement of the stake was measured at two heights, about 13 cm (5 in.) and 50 cm (20 in.) above ground level, with dial gages. The upper gage had a maximum travel of 15 cm (6 in.) while the lower had only a 2.5 cm (1 in.) travel. Both were graduated in 0.0025 mm (0.001 in.) divisions.

Each test was performed by first driving a stake to the prescribed depth with a 3.6 km (8 lb) sledgehammer, adjusting the loading frame to contact the stake, and setting up the dial gage frame. Figure 6 shows a test ready to be performed. Loads were applied in increments of about 45 N (10 lb) after which dial readings were made. For increments up to 450 N (100 lb), the load

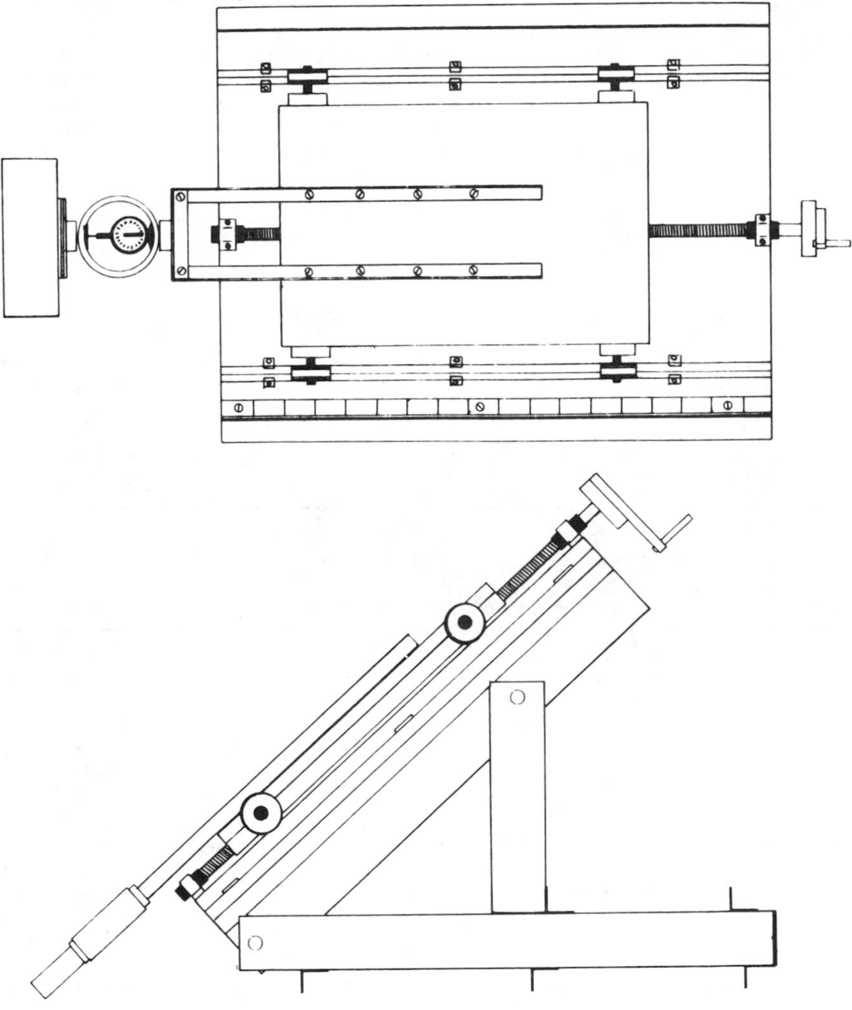

FIG. 5—*Loading mechanism.*

was reduced to zero after each increasing increment to observe cyclic loading behavior. For loads greater than 450 N, the load was applied monotonically.

Results

A total of 139 lateral stake tests was performed. The major variables studied were: stake size, stake length, angle of load, angle of stake, stake-to-load connection, and soil type. Stake size and length were major factors in measured lateral capacity. It was found that the angle of the load could be adequately accounted for by the horizontal component only, and the angle of the

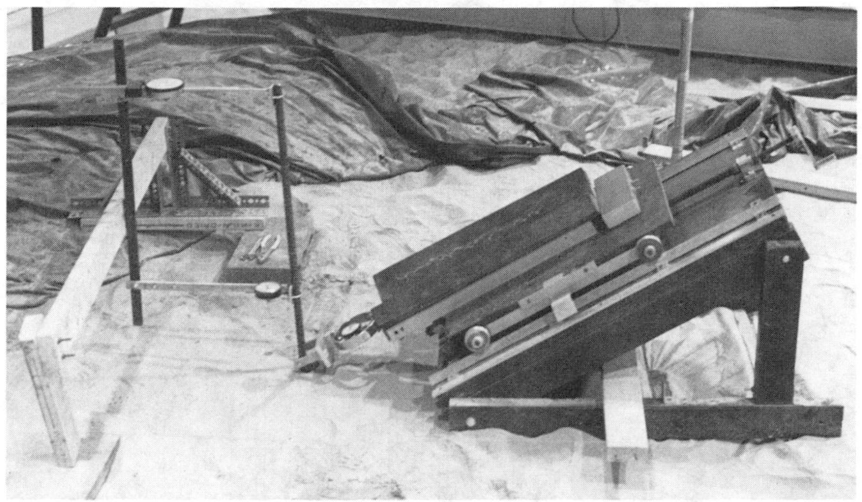

FIG. 6—*Ready to test.*

stake could be accounted for by the stake length projected on a vertical plane. The stake-to-load connection had only a minor influence on the lateral capacity. However, there were indications that the collar connection was introducing a rotation which had an undesirable effect, and this connection was abandoned early in the program.

One set of tests was performed by placing a greased sheet of plywood on the ground under the 3.8 by 8.9 by 25 cm (2 by 4 by 10 in.) load connection to evaluate the influence of ground friction on lateral stake capacity. Comparing results of these tests with those without the greased plywood showed no difference in lateral stake capacity.

Figure 7 shows typical load-displacement curves for one soil and three stakes. Each curve is the average of three tests. All tests in the program showed this same general shape, that is, continuous curvature, no yield, and no sharp failure. It was necessary, therefore, to select some criterion for comparing results and formulating a mathematical model for stake capacity. We decided the tests would be evaluated at two displacements, 3.2 mm (1/8 in.) and 6.4 mm (1/4 in.). Displacements of stakes on both sides of a masonry wall could then amount to 6.4 and 12.8 mm (1/4 and 1/2 in.), respectively. Concerns over cyclic loading which is accompanied by accumulative permanent displacement lead to the selection of 3.2 mm (1/8 in.) as the allowable displacement. Measured horizontal loads for 3.2 mm (1/8 in.) ground level displacements for three types of stakes of two depths and angles of loading are shown in Table 3.

All test data at displacements of 3.2 mm (1/8 in.) were compared with an equation for lateral pile capacity based on Terzaghi's theory of subgrade reac-

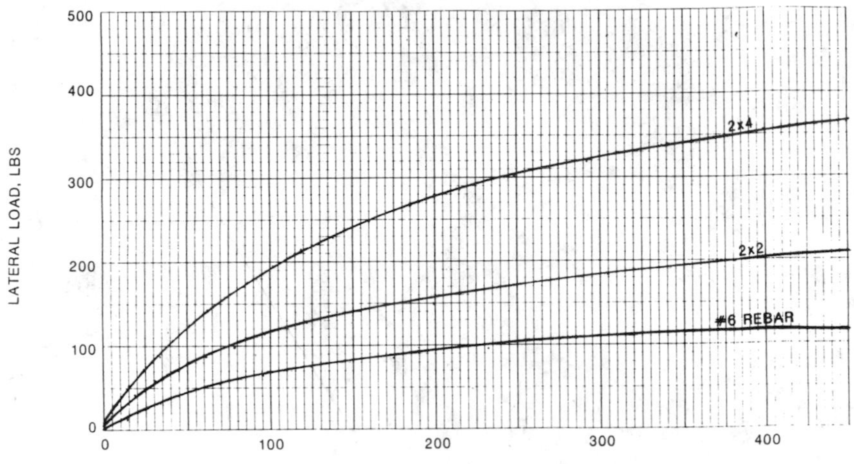

FIG. 7—*Load-displacement curve.*

tion [*1*]. This equation by Barber [*2*] assumes linear elastic behavior of the soil and rigid behavior of the stake. For the 45 cm (18 in.) long stakes of No. 6 rebar and 3.8 by 3.8 cm (2 by 2 in.) this is a good working assumption but is at the limit based on stake stiffness. However, for shorter embedment it is a good assumption for all stakes. To obtain a comparison between tests, the load required to produce 3.2 mm (1/8 in.) ground surface displacement in the experiment was plugged into the Barber equations to calculate an "effective" coefficient of subgrade reaction. After observing the general trends obtained in this way, it became evident that better correlation could be obtained with a slight modification to the Barber equation. The modification involved only the depth to the point of stake rotation and yielded the following simple equations:

Granular soils (sand)

$$H = \frac{y_0 n_h L^2}{10} W \tag{1}$$

Cohesive soils (clay)

$$H = \frac{y_0 k_h L}{4} W \tag{2}$$

TABLE 3—*Sample data, horizontal load* H *in pounds (N) needed for a deflection of 0.125 in. (3.2 mm).*

Load Angle, deg	Embedment	Stake	ML	SP	CL—Stiff	CL—Medium	CL—Soft
			H	*H*	*H*	*H*	*H*
30	18 in. (45 cm)	No. 6 rebar (1.9 cm)	71 (318)	75 (336)		126 (565)	
		2 by 2 (3.8 by 3.8 cm)	142 (636)	128 (574)		120 (538)	
		2 by 4 (3.8 by 8.9 cm)	243 (1089)	219 (982)			
45	18 in. (45 cm)	No. 6 rebar (1.9 cm)		75 (336)			
		2 by 2 (3.8 by 3.8 cm)		116 (520)			
		2 by 4 (3.8 by 8.9 cm)	271 (1215)	198 (887)			
30	12 in. (30 cm)	No. 6 rebar (1.9 cm)	50 (224)	57 (255)			
		2 by 2 (3.8 by 3.8 cm)	99 (444)	66 (296)			
		2 by 4 (3.8 by 8.9 cm)	156 (699)	107 (480)			
45	12 in. (30 cm)	No. 6 rebar (1.9 cm)	46 (206)	67 (300)	203 (910)		96 (430)
		2 by 2 (3.8 by 3.8 cm)	97 (435)	104 (466)	302 (1354)		83 (372)
		2 by 4 (3.8 by 8.9 cm)	145 (650)	315 (1412)	315 (1412)	185 (829)	101 (453)

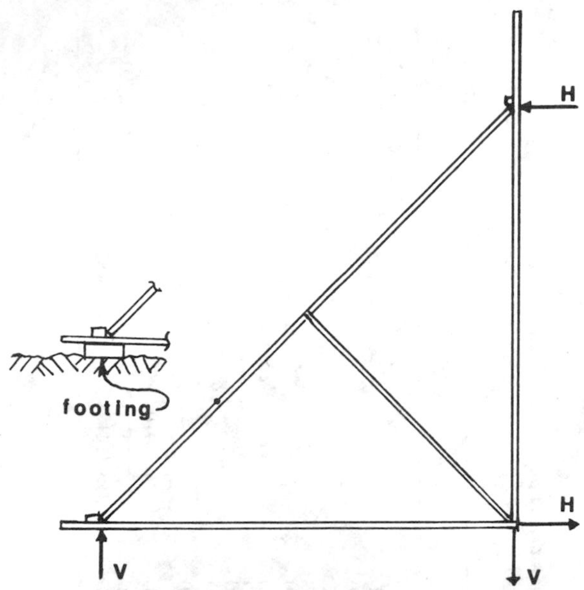

FIG. 8—*Truss-type brace.*

where

H = horizontal component of load (1 lb = 4.5 N),
W = width of stake (1 in. = 2.54 cm),
y_0 = stake displacement at ground level (1 in. = 2.54 cm),
n_h = constant of horizontal subgrade reaction (sand),
k_h = coefficient of horizontal subgrade reaction (clay), and
L = embedded length of stake (1 in. = 2.54 cm).

The coefficients of subgrade reaction back-calculated in this way correlate well with those presented by Teng [*3*].

Equation 1 assumes the depth to the point of rotation of the stake is $5/9\,L$ while the Barber equation uses $3/4\,L$. For a specific test the predicted horizontal load using Eq. 1 was 1.03 kN (230 lb) while the measured value was 1.01 kN (225 lb).

These simple equations, 1 and 2, can provide a reasonable estimate of the horizontal capacity of stakes if the soil is correctly identified. However, in most soils to be found at the surface of the ground at a construction site the stake capacity will be less than 1.34 kN (300 lb). This capacity is not sufficient to support a reasonable length of a tall masonry wall even in a moderate wind, since braces would be spaced at about 1.3 m (4 ft) for a 56 km/h (35 mph) wind on a 6.1 m (20 ft) wall.

Suggested Alternative

With soils, vertical resistance can be more easily achieved than horizontal resistance. The addition of a horizontal tension member will form a truss, as shown in Fig. 8, which transmits the force H to the foundation. Only a vertical force V, equal to H, is required at the end of the truss. Any soil which will support a man standing on one foot will carry 4.8 kN/m^2 (1000 psf). Temporary footings are easily provided by one or two concrete blocks. The lumber in the truss becomes the limiting factor in the capacity of the shoring. Spacing between trusses can be increased to a practical distance of 4.5 to 6 m (15 to 20 ft), depending on the wood used. The detail at the bottom intersection of truss and wall is particularly important, because the connection must transmit the force tension $H = V$ both horizontally and vertically between the truss and the wall. The truss type of configuration follows that of steel bracing systems, but uses materials common to the job site.

Acknowledgments

This research was supported by a Safety Education and Training grant from the Bureau of Safety and Regulation, Michigan Department of Labor. We wish to thank the Department of Labor, the Advisory Committee, and the Masonry Institute of Michigan for their help.

References

[*1*] Terzaghi, K., "Evaluation of Coefficient of Subgrade Reaction," *Geotechnique*, Vol. 5, 1955, p. 297.

[*2*] Barber, E. S., discussion of paper by S. M. Gleser, *ASTM STP 154*, American Society for Testing and Materials, Philadelphia, 1953, pp. 96–99.

[*3*] Teng, W. C., *Foundation Design*, Prentice-Hall Inc., Englewood Cliffs, NJ, 1962.

Walter M. Naish[1]

A Study of Glazed Brick Durability Characteristics and Building Applications

REFERENCE: Naish, W. M., **"A Study of Glazed Brick Durability Characteristics and Building Applications,"** *Masonry: Research, Application, and Problems, ASTM STP 871*, J. C. Grogan and J. T. Conway, Eds., American Society for Testing and Materials, Philadelphia, 1985, pp. 126–137.

ABSTRACT: The author's company has been manufacturing glazed brick for more than 50 years. From a review of quality control and independent testing laboratory records, it is evident that this product has consistently met ASTM Specification for Facing Brick (Solid Masonry Units Made from Clay or Shale) (C 216) standards for durability. In early 1981 the author initiated a research study to determine some of the characteristics used by architects and structural engineers in designing the exterior walls of a variety of successful buildings utilizing these glazed brick.

The data gathered demonstrate that glazed brick with durability characteristics such as those reported have been used (1) in exterior building walls having a variety of design criteria, (2) in buildings of differing sizes and end uses, and (3) in severe weather climates. The author goes on to discuss that some wall failures have occurred in buildings other than those reviewed in this study; to comment regarding his observations of some such failures; and to suggest ongoing research aimed at more clearly defining in-service conditions that contribute to wall failures.

KEY WORDS: clays, masonry, brick, building information

Much research has been done and is still in progress regarding the durability of clay masonry products and the building environments in which they are utilized. The wide variety of manufacturing circumstances on one hand, and of building applications on the other, have made it very difficult to draw hard and fast conclusions.

The study presented here is not based on research done in a laboratory environment, but rather on a review of historical durability characteristics from both a product and a building application point of view. Hopefully, it will

[1]President, Hanley Brick Inc., Summerville, PA 15864.

make a meaningful contribution to our industry knowledge and will point the way to additional future research.

The author's company has been manufacturing a variety of clay products from the same clay formation at its Summerville tunnel kiln plant since it was originally built in 1926. Glazed brick, a part of this product line, has been produced since about 1928.

Through the years, an integral part of manufacturing operations has been a well planned, in-house, quality control (QC) program. Its practices and procedures include testing of raw materials, monitoring processes, and testing of both ware-in-process and finished products. In addition, finished products are sent out frequently for testing by independent testing laboratories. Some of these latter tests are initiated as routine checks by the company; some are customer-requested.

A review of company records by the author indicated that a wealth of information was available. Included in the record files were test data associated with product durability—24-h cold submersion absorption, 5-h boiling absorption, saturation coefficient, initial rate of absorption, compressive strength, and freeze/thaw tests.

With the availability of this extensive background of product quality data, it seemed to the author that an "end-use" study of buildings which utilized these glazed brick and which were performing satisfactorily might yield additional useful information. Accordingly, a study was planned to target on gathering data about the exterior wall design, construction, and maintenance factors incorporated in a variety of successful building applications. Both QC record analysis and building fieldwork were launched in early 1981.

This study has been directed towards fact-finding only. It is not intended to recommend or promote specific building design, construction, or maintenance techniques.

Procedure

Product Quality Information

In-House Sampling and Testing—Fired ware is sampled directly from the kiln cars generally twice a week at about the same hour. Three full-brick specimens are taken for absorption testwork from specific car locations to identify the maximum range of absorptions that might occur across the glazed brick load. Depending on production schedules, glazed brick may be sampled up to 13 times per month (average = 6 times).

Nonglazed face brick are sampled using the same procedures on an average of 14 additional times per month. Other structural and industrial products are also regularly checked in a similar fashion. However, only glazed brick data are reported in this study.

Independent Laboratory Testing—Glazed brick specimens for this testwork are usually randomly sampled from the finished goods inventory. Occasionally selective sampling is done to spotlight a specific factor. Independent laboratory testwork may include only one or two particular tests (e.g., freeze/thaw); or a complete test series might be ordered.

Building Information

Buildings which were erected in the period 1950 to 1970 and which are performing satisfactorily were identified by in-company or distributor sales management personnel. A two-page fact sheet was prepared. One page covered exterior, visual building characteristics and was completed by in-company personnel. The other page covered some wall design characteristics and was completed by an independent, knowledgeable, third party, such as an architect, specifier, or maintenance manager. Building locations are all in the North Central US, North East US, or in Canada.

Additional Information

As the research program described above unfolded, other facts and testwork related to product durability or satisfactory building performance were identified. These include observations of buildings dating back 40 to 50 years, and observations of five buildings on which glaze crazing had occurred. A shipping/order record search was also made to determine the approximate quantities of glazed brick that have been manufactured and shipped through the years.

In the spring of 1982, it was decided to add to the project a special freeze/thaw test program. It had two purposes: one was to test specimens to failure; the other was to determine the effect of freeze/thaw action on crazed glazes. Two lots of glazed brick specimens were randomly selected—one from inventory about 10 years old; the other from current stock. They were subjected to routine, in-house tests which indicated that 24-h cold and 5-h boiling absorptions as well as the saturation coefficient ratios were within "normal" ranges. Additionally, they were autoclaved according to standard test methods and did not craze. Other specimens were subjected to longer-than-normal autoclaving until "forced" crazing did occur. By judgment, individual units were ranked as having light, medium, or heavy crazing. Noncrazed and crazed specimens were sent to an independent testing laboratory for freeze/thaw testing.

Experimental Work and Results

Product Quality Information

In-House Data—Statistics resulting from tests of 24 h cold submersion absorption, 5 h boiling absorption, and initial rate of absorption as well as saturation coefficient calculations have been averaged for those years in which meaningful records could be found. This information is shown in Fig. 1.

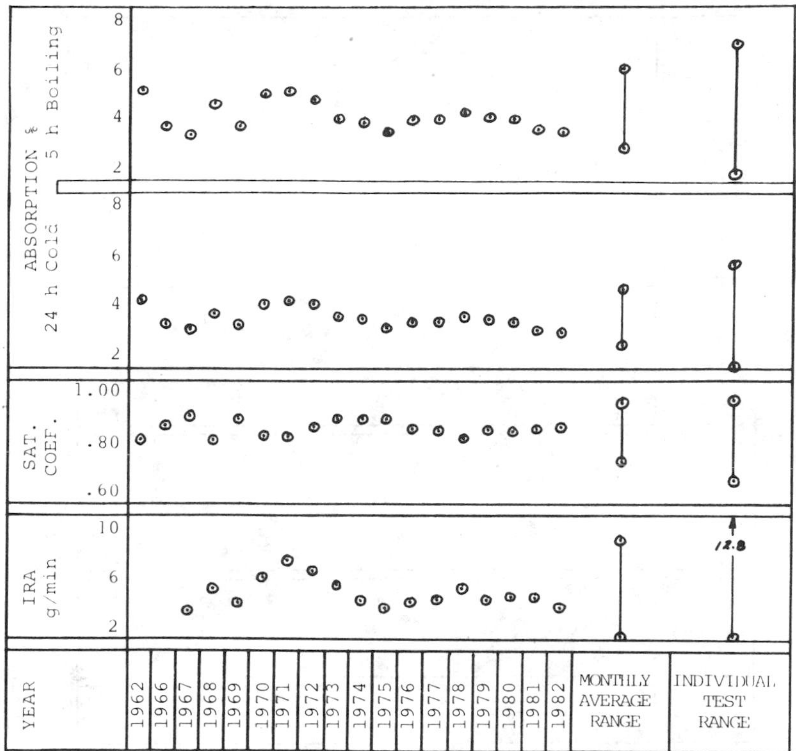

FIG. 1—*Monthly average and individual test ranges, annual average results (in-house data).*

Independent Laboratory Data—Compressive strength, 24-h cold submersion, and 5-h boiling absorptions and saturation coefficient calculations of individual tests are shown in Fig. 2, for those years in which data could be located.

In addition to the special program started in 1982, at least five freeze/thaw tests have been done through the years, the earliest on record being in 1955.

Building Information

Data gathered from 18 buildings by in-company personnel are presented in Tables 1 and 2. Additional data and commentary gathered from the same 18 buildings by independent sources are shown in Tables 3 and 4.

Additional Information

Three buildings erected in the early Thirties and Forties were identified and examined by in-company personnel. Because it was normal in that era, it is

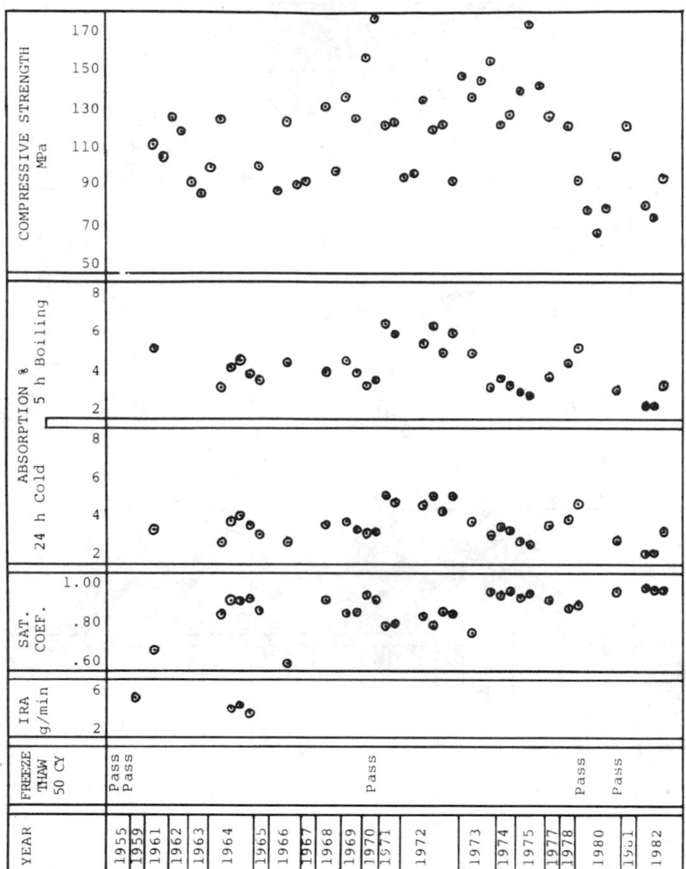

FIG. 2—*Individual test results (independent testing laboratory data).*

presumed that they are all of solid wall construction—though no design detail was available to support this. Each appeared to be well maintained and was performing satisfactorily.

Coincidental with the search of records, five buildings were identified on which glaze crazing had developed while the brick were in service. Each of these crazed wall areas has been inspected and found to be in satisfactory condition. Two were erected in the Forties, two in the Fifties, one in the Sixties.

Order/shipping records have been examined and indicate that more than 11 000 orders have been processed, resulting in shipments of at least 370 000 000 glazed brick. It is believed that the vast majority of these were used in exterior, severe weathering applications, though no information is available to provide an exact breakdown.

The freeze/thaw test program which was started in the spring of 1982 is still going on. All units—noncrazed and crazed, those from 10 year inventory, and those from current stock—have passed 350 cycles without any failure.

TABLE 1—*Building data by Hanley personnel.*

City	Pittsburgh PA	Johnstown PA	Pittsburgh PA	Pittsburgh PA	Mount Lebanon (Pgh) PA	New York NY	New York NY	Verdun, Que. (Montreal)	Montreal Quebec
Year	1970	1966-67	1958	1968	1965	1956	1956	1962	1957 & 1959
Building use	2 bldgs. sewage	service & maintenance	office	water treatment	office	office	convention ctr & off.	hospital	apartment Phases I & II
Number floors	2 & 6	2	2 & 4	3 & 4	10 + 2 parking	34	34	2	12 +
Brick—Quantity	375M	87M	75M	160M	450M	500M	700M	45M	1500M
Brick—size	standard	standard	standard	norman	standard	standard	standard	standard	standard
Parapet design	yes	partial	yes	yes	yes	yes	yes	yes	yes
Coping used	metal	metal	limestone	granite	metal & limestone	metal	metal	metal	metal
Overhanging roof	no	partial	no	no	no	no	no	no	yes
Mortar joints	concave tooled	concave tooled	concave tooled	concave tooled	concave tooled	concave tooled	concave tooled	concave tooled	concave tooled
Expansion joints	vertical	vertical	none	none	vertical	horizontal	vertical	nil	vertical
Weep holes	yes	no	no	no	no	no	no	no	yes on Phase II
Upper wall vents	no	no	no	no	no	no	yes	no	no
Drip edges	yes	yes	yes	yes—windows no—coping	yes	yes		yes	yes
Brick at grade	yes	yes	yes	yes	yes	no	no	yes	no
Opening sills	metal	stone	metal	metal	metal	metal	metal		metal

TABLE 2—*Building data by Hanley personnel (continued).*

City	Montreal Quebec	Cleveland OH	Syracuse NY	Utica NY	Sheridan Park Ontario	Etobicoke Ontario	Chicago IL	Chicago IL	Wheeling IL
Year	1955	1964	orig—1960 add—1973	1967	1956	1956	1949-50	1967	1967
Building use	commercial & office	office	hospital	library & office	research	commercial & shopping plaza	automobile dealer	automobile dealer	office & warehouse
Number floors	10	4	6	4	2	2	4	1	1
Brick—quantity	175M	100M	orig—500M add—312M	344M	80M	150M	200M	86M	240M
Brick—Size	standard	standard	standard	standard	standard	standard	standard	standard	standard
Parapet design	yes	no	no	no	no	yes	yes	partially	no
Coping used	metal	metal on penthouse	6-in. copper over wall at roofline		metal	brick	concrete & stone	metal	metal
Overhanging roof	yes	yes	no	yes	no		no	partially	no
Mortar joints	concave tooled	concave tooled	concave tooled		concave tooled	concave tooled	raked slightly	concave tooled	concave tooled
Expansion joints	nil		vertical		vertical	vertical	horizontal	vertical	vertical
Weep holes	no	no	no	no	no	yes	no	yes	no
Upper wall vents	no	no		no	no		no	no	no
Drip edges	yes	no	no	yes	yes	yes	yes	yes	yes
Brick at grade	no	yes	no	yes	no		no	yes	yes
Opening sills	metal	metal	limestone	metal	metal		stone	stone	

TABLE 3—*Building data by independent personnel.*

City	Pittsburgh PA	Johnstown PA	Pittsburgh PA	Pittsburgh PA	Mount Lebanon (Pgh) PA	New York NY	New York NY	Verdun, Que. (Montreal)	Montreal Quebec
Year	1970	1966–67	1958	1968	1965	1956	1956	1962	1957 & 1959
Wall design	solid	cavity	solid	3 Wythe cavity	cavity	solid	solid	3 Wythe solid	cavity
Backup material	clay brick	concrete block	concrete block	structural & glazed tile	concrete block	clay brick	concrete block	concrete block	concrete block
Insulation	no	styrofoam	plaster on metal lath	1-in. rigid	yes	no	no	styrofoam	fiberglass
Vapor barrier	no	no	2 coats mastic bondex	no	no	no	troweled mastic	no	fiberglass backing
Building frame	steel	steel	steel: concrete on 2 floors below grade	steel	steel	steel	concrete	reinforced concrete block	concrete
Flashing—									
at roof/parapet	yes		yes	stainless steel	yes		yes		yes
over openings	yes		yes	yes	yes		yes	yes	yes
under 1st course	yes	yes	yes		yes			yes	yes
Mortar specs.	cement/lime	masonry mortar	cement/lime & masonry mortar	cement/lime	cement/lime	cement/lime	cement/lime type N 1:1:6	colored	colored
Comments	(1)	(2)	(3)	(4)	(5)	(6)	(7)	(8)	(9)

(1) Buildings in excellent condition—well maintained.
(2) Building in good condition—slight spalling under 2 concrete stoops—maintenance normal.
(3) Building in good condition—well maintained—some structural cracking repaired—addition 1979.
(4) Building in excellent condition—well maintained—cleaned 2 or 3 years ago—addition 1980.
(5) Building in excellent condition—some crazing near roof—good maintenance—inspected every 2 or 3 years—repointed as needed.
(6) Maintenance—visual inspection annually. Glazed brick tied to backing with stainless steel wire ties.
(7) Maintenance—visual inspection approximately every six months—structural cracks patched with soft joints—GB parged on back.
(8) 4-in. GB, 1-in. air space, 8-in. concrete block, insulation, space, 4-in. concrete block. Maintenance—no problem.
(9) Half a dozen spalling at flower boxes. One horizontal crack on Phase I—a couple of settlement cracks at floor levels.

TABLE 4—*Building data by independent personnel (continued).*

City	Montreal Quebec	Cleveland OH	Syracuse NY	Utica NY	Sheridan Park Ontario	Etobicoke Ontario	Chicago IL	Chicago IL	Wheeling IL
Year	1955	1964	orig—1960 Add—1973	1967	1956	1956	1949–50	1967	1967
Wall design	solid	solid	solid	solid	solid	solid & cavity	solid	solid	solid
Backup material	concrete block	concrete block	concrete block	concrete block	concrete block	concrete block	concrete block	concrete block	concrete block & glazed tile
Insulation	cork	½-in. rigid	no	no	no	yes styrene/ glass fiber	granular in block cells	granular may be in block cells	vermiculite & rigid
Vapor barrier	no	foil-back drywall	no	no	no	yes			no
Building frame	concrete	steel	concrete	steel	steel	steel	concrete	steel	steel
Flashing—									
at roof/parapet	yes	yes	yes	yes	yes	under coping & openings	yes	yes	
over openings	yes		yes	yes	yes		yes	yes	yes
under 1st course						yes			yes
Mortar specs.	colored	cement/ lime	cement/ lime	cement/ lime	cement/ lime	stoneset or colored or both			cement/lime masonry mortar
Comments	(10)	(11)	(12)	(13)	(14)	(15)	(16)	(17)	(18)

(10) Walls constructed with glazed brick, air space and backed with 8-in. cement block—allows breathing.
(11) Building looks good—large overhang roof—no large surfaces.
(12) Addition match—perfect—maintenance program for exterior and interior.
(13) Building in excellent condition—good maintenance program.
(14) Building maintenance-free due to brick and workmanship.
(15) Good quality brick—excellent appearance, very little maintenance required.
(16) Building appearance/masonry work in reasonably good condition. Slight crazing and settling cracks seem associated with building movement and backup material.
(17) Building masonry in excellent condition—structure sound—only one crack maybe due to building movement.
(18) Building appearance excellent—recently tuckpointed—free-standing sign base deteriorated (no coping overhang, coping cracks, holes to accommodate sign).

Conclusions

A study of the data presented herewith leads to two primary conclusions:

1. Both the historical and current ASTM Specification for Facing Brick (C 216) durability criteria for absorption and freeze/thaw resistance have been met without exception. Not only were all annual and monthly absorption averages well below 8%, but no single 24-h cold or 5-h boiling test was observed to be over 8% (the maximum 24-h cold absorption was in fact 5.8%). Freeze/thaw tests throughout the years have always passed the 50 cycle standard and, as the current test (at 350 cycles and still counting) shows, have far exceeded it. Initial rate of absorption and compressive strength results have also been well within the recommended limits.

The saturation coefficient ratio, on the other hand, has exceeded 0.78 on all annual averages, though a few monthly averages were at or below that level.

Glaze crazing appears not to be a negative factor except only perhaps in the aesthetic sense.

2. The exterior walls of buildings utilizing these glazed brick have been designed and constructed with a wide variety of detail, and in the cases studied, have performed satisfactorily for many decades in severe weather climates. It seems evident from this review that if design, construction, and maintenance practices are satisfactory, a great deal of design freedom is practical.

Discussion

When one considers the high compressive strength which is characteristic of these brick, it is difficult to imagine that they could fail due to in-the-wall compressive loading.

Also, when one considers their low 24-h cold absorption characteristics, as well as the excellent freeze/thaw cycle performance, it is difficult to imagine their failure because of water penetration into the wall system.

Again, when one considers the many examples of satisfactory exterior building wall performance using a wide variety of design characteristics over a period of many decades, it is difficult to imagine that brick meeting ASTM C 216 durability specifications as consistently as these could fail in service.

Yet, in recent years a small percentage of walls in buildings not reported in this study, but utilizing these glazed brick, have failed. Why?

It is the author's conclusion that the wall failures he has observed are not the result of (*a*) too generous ASTM specifications or (*b*) inconsistent brick quality, but rather of excessively hostile wall conditions into which the brick have been placed.

As an example, is it reasonable to expect brick (even those with compressive strengths ranging upward from 9000 psi) to withstand the uncontrolled compressive load of a shrinking concrete framed building which has been designed with inadequate horizontal expansion joints? Walls utilizing these bricks have done just that in many cases, I believe—but sometimes the loading is so severe that the wall cannot take it, and failure is exhibited, generally in a well defined pattern.

Again as an example, is it reasonable to expect brick (even those with absorption characteristics, freeze/thaw performance, and an in-wall track record as well proven as these) to withstand wall conditions which encourage water to literally drain into the brick wythe? The author's observations suggest that most wall areas subjected to such punishment do, in fact, withstand it for extended periods; but on some occasions, unfortunately, water entry and freeze/thaw conditions are so severe and long lasting that wall failure does occur—first as cracking, then sometimes as spalling.

Virtually all wall distress resulting in spalling that the author has observed occurs in identifiable areas and displays a more or less consistent pattern: i.e., an area with darkened (wet or stained) mortar joints, within which one frequently may observe (*a*) efflorescence, (*b*) a number of cracked brick and mortar joints, and (*c*) a lesser number of spalled brick and mortar joints. Additional observations nearly always point to a water source associated with the "identifiable area."

The author's observations also indicate that the early signs of a wall problem are darkened (wet or stained) mortar joints and efflorescence. If cracked and then spalled brick and mortar joints subsequently occur, it is usually much later. At the darkened mortar joint stage, problems can be and frequently are relieved by means of appropriate maintenance.

Another distressed wall situation the author has observed is that of water migrating considerable distances within a wall, then becoming entrapped and retained to form pools. Such pools may occur in brick cores, in partially filled head joints, and likely in partially filled collar joints. If they become frozen, a brick face or wall area may be cracked or spalled after only a relatively few freeze/thaw cycles. A more insidious effect, one suggests, is that the ice expansion may be partially contained by the strength and rigidity of the wall assembly, but that some tensile strength loss may occur, microcracks may form, and mortar-to-brick bonds may be disrupted. Even a slight occurrence of any of these effects may allow the next influx of water to reach further into the components of the wall assembly, thus leading to a progressive and more extensive type of wall deterioration.

Suggested Future Research

Through the years, a great deal of research aimed at improving masonry wall durability has been done, and is still going on. This includes studies of all wall component characteristics and their interaction one with another, as well as the performance of walls in situ. Yet more can be done.

The work and conclusions described herein suggest that additional future research should focus on water penetration and freeze/thaw characteristics related to walls which have performed well and those which are at various stages of distress—particularly the early, darkened-joint stage.

Two approaches might be taken:

1. Field research to determine the conditions existing in satisfactory and distressed walls; i.e., water sources, water migration, water retention, climatic conditions, freeze/thaw occurrences, wall design-workmanship-maintenance characteristics, quality and compatibility of wall components, and probably others.
2. Laboratory research in which a variety of wall sections are exposed to various controlled, measurable, hostile conditions.

From such research in the field and in the laboratory, conclusions might be drawn which would better recognize early signs of potential wall problems, component limitations, component interactions, and climatic impact; which in turn might lead to wall design, workmanship, and building maintenance criteria more closely geared to improved masonry wall life.

Arie Huizer[1] *and Michael A. Ward*[1]

Thermal Shock and Maximum Temperature Investigation of a Single Unit Clay Brick Masonry Chimney

REFERENCE: Huizer, A. and Ward, M. A., **"Thermal Shock and Maximum Temperature Investigation of a Single Unit Clay Brick Masonry Chimney,"** *Masonry: Research, Application, and Problems, ASTM STP 871*, J. C. Grogan and J. T. Conway, Eds., American Society for Testing and Materials, Philadelphia, 1985, pp. 138–150.

ABSTRACT: A double-wall clay chimney flue liner, which has been shown to pass the requirements for "zero clearance" applications at hearth temperatures of approximately 700°C (1292°F), was subjected to an additional series of tests presently only prescribed for metal chimneys. Although the National Building Code of Canada at this point in time does not require clay flue liners to pass the test requirements set for metal chimneys, the authors considered it to be a distinct possibility that in the near future there will be some pressure to have clay liners attain the same standards that the metal ones presently are required to meet. Therefore, the unit under investigation was subjected to the most severe conditions that might be experienced by clay flue liners such as thermal shock, creosote burnout, and extremely high flue gas temperatures. Test results showed that at present the unit would not pass the new standards set for metal chimneys, and that it will be necessary to reevaluate its geometry as well as the properties of the clays used in its manufacture.

KEY WORDS: masonry, chimneys, flue liners, thermal shock, burnout test

Because of the requirements resulting from the advances made in the design of commercial and residential heating equipment, in addition to the increasing usage of liquid and gaseous fuels, the proven and architecturally attractive masonry chimney has undergone some design changes which have increased rather than solved its problems. The traditional flue with its large cross-sectional area permitted effective flue gas cooling and minimized the requirement for soot cleaning. However, the higher draft requirements

[1]Senior instructor and professor and head, respectively, Department of Civil Engineering, The University of Calgary, Calgary, Alberta, Canada.

demanded by more modern equipment resulted in the reduction of cross-sectional areas of flue liners while at the same time increasing flue gas temperatures. Also, to reduce its demand on area, masonry surrounding the flue was minimized, increasing the possibility of moisture penetration and susceptability to freeze-thaw effects. Such problems, in addition to poor workmanship, have resulted in more rapidly decaying joints and increased loss of heat to the surrounding area.

The authors have previously reported on the development of a clay unit [*1,2*] which would solve most of the problems encountered, Fig. 1. The main competitor of the clay liner, the metal chimney, has been the object of an extensive investigation. The improvements which have resulted from these investigations have produced the minimum design for a metal chimney and are reflected in the specifications laid down in Underwriters Laboratories Standard for Factory Built Type A Chimneys (ULC-S629-M1981).

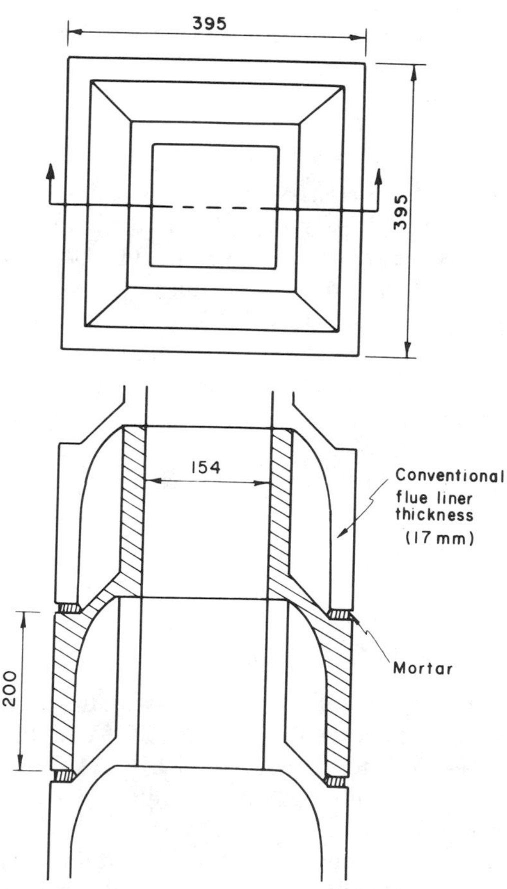

FIG. 1—*Unit tested.*

The necessity of meeting these requirements has increased the price of metal chimneys appreciably, putting the clay liner in a more favorable position. But it is probably somewhat naive not to expect pressure on building code authorities to require manufacturers of clay liners to meet similar stringent requirements in the future. This consideration prompted the authors to initiate a series of tests to evaluate the new unit at a level comparable to that expected of metal chimneys.

Testing Program

In accordance with the specifications outlined in ULC-S629, a flue gas generator was constructed with chrome-alumel thermocouples located at its flue pipe. The test assembly is shown in Figs. 2 and 3. Figure 4 shows the location of the thermocouples, copper-constantan for the surface and iron-constantan for the interior.

Four of the tests that were considered most relevant to clay brick chimneys were selected for this program from the standards laid down for metal chimneys.

Test No. 1—Thermal Shock

The temperature of the flue gases entering the chimney was regulated by the quantity of primary and secondary air induced into the generator. The flue gas generator was fired at the input specified by ULC-S629-M1981 for a chimney with a nominal diameter of 152 mm.

At the specified input and by adjusting the damper on the flue gas generator, the temperature of the flue gases as measured by the thermocouple grid was kept at 900°C (1620°F) above room temperature. The test was continued for 10 min at this rate of heat input and flue gas temperature, at which point the generator was shut off.

The test was repeated a further two times with a 4-h cool down period between each test.

No surface temperature readings were taken, but at the conclusion of Test 1 the chimney was visually inspected.

Test No. 2—Temperature—650°C Flue Gases

This test started with the chimney at room temperature. The flue gas generator is fired at the specified input given in ULC-S629 and the air supply regulated to give flue gases at a temperature of 625°C (1125°F) above the room temperature. The test was continued until equilibrium temperatures were attained on the surfaces of the chimney.

For zero clearance it is a condition of this test that those parts of the chimney at points of zero clearance to the structure shall not be more than 65°C (117°F) above room temperature either during or immediately after this test.

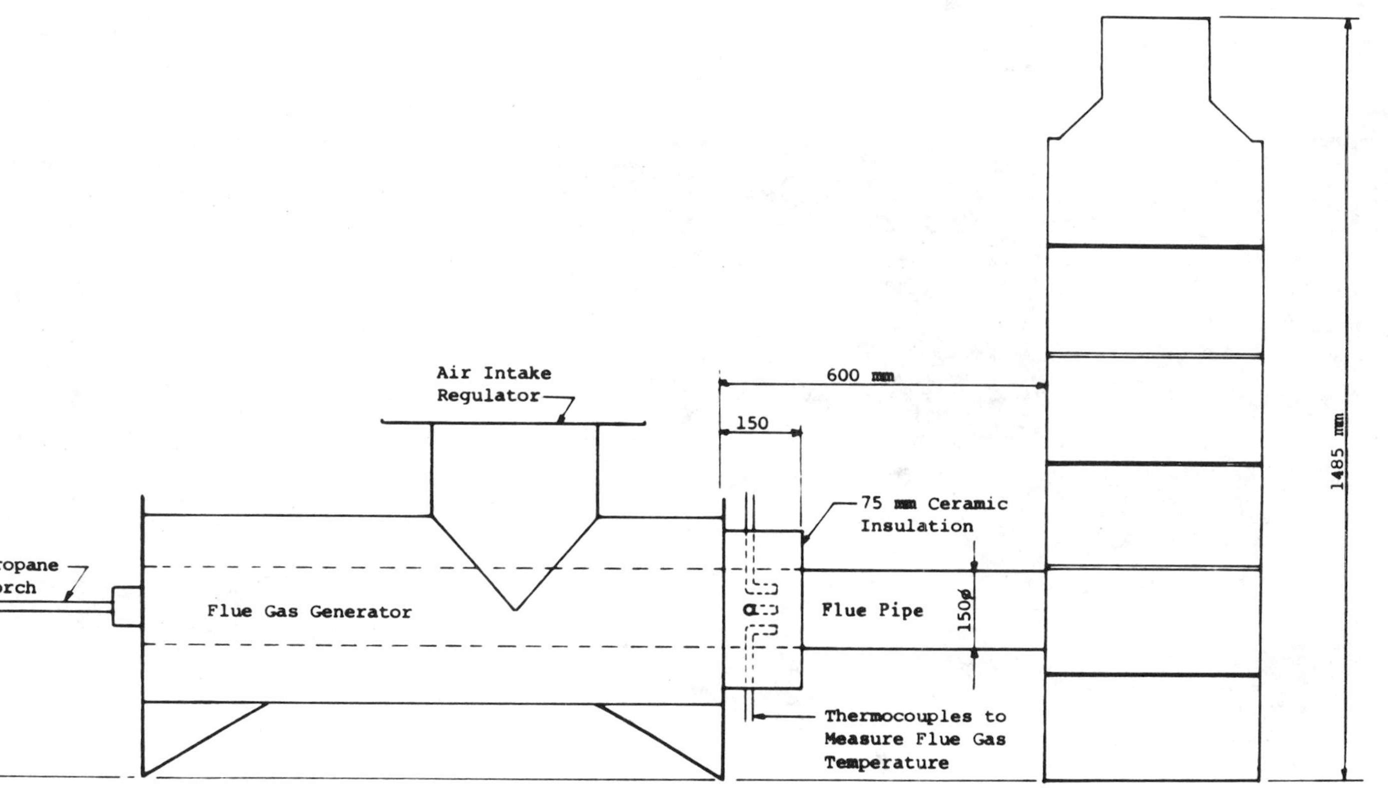

FIG. 2—*Schematic of test assembly.*

FIG. 3—*Test assembly.*

Test No. 3—Temperature—925°C Flue Gases

After the equilibrium conditions were attained under the test conditions given in Test 2, the flue gas generator was fired to an increased input to give flue gases at a temperature of 900°C (1620°F) above room temperature and continued for 1 h. The requirements of this test are that the surfaces of the test chimney at points of zero clearance to the structure should not be more than 78°C (140°F) above room temperature. The room temperature was the temperature as measured at the end of the 1-h firing period.

Test No. 4—Temperature—Creosote Burnout

This test was started with the chimney at room temperature. The flue gas generator was fired at a specified input to produce flue gases at a temperature of 300°C (540°F) above room temperature. The generator was fired in this manner until equilibrium temperatures were attained on the surface of the

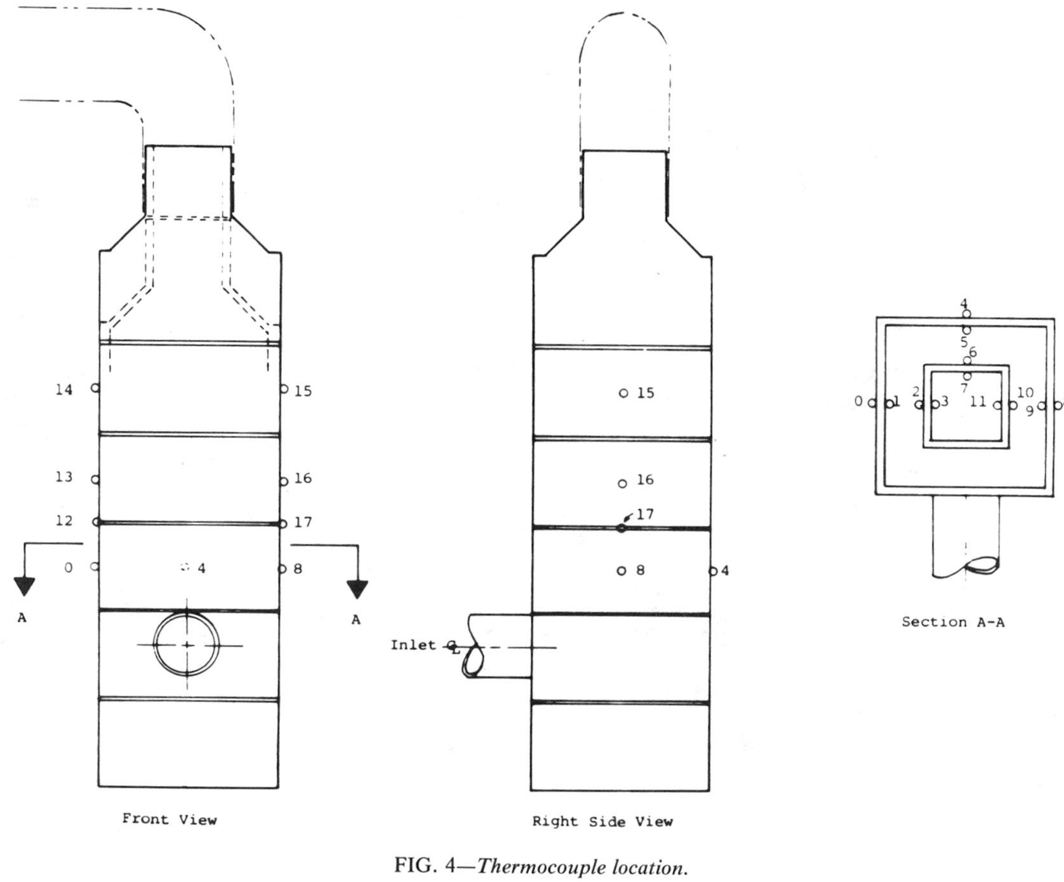

FIG. 4—*Thermocouple location.*

chimney. After equilibrium temperatures were attained, the input to the flue-gas generator was increased and the air supply regulated to produce flue gases at a temperature of 1125°C (2025°F) above room temperature. This rate of input was maintained for 30 min and then the generator was shut off. This test was repeated for a total of three cycles. The first cycle was conducted with the connector pipe from the generator to the chimney uninsulated for a length of at least 450 mm from the thimble. The two remaining test cycles were conducted with the full length of the connector pipe insulated with a 75 mm layer of ceramic blanket insulation. The requirements of the test were that the surface of the chimney at points of zero clearance to the structure should not be more than 97°C (175°F) above room temperature during the firing nor after the generator was shut off. The room temperature was the temperature as measured at the end of the 30 min firing period.

Results

Test No. 1—Thermal Shock

The following observations were made:

(*a*) The chimney showed no sign of cracks forming in the flue liners themselves.

(*b*) Due to differential movement between the inner and outer liner, there were signs of the mortar joints on the outer liner opening up.

(*c*) The joint with the metal thimble immediately below it showed the greatest separation.

(*d*) All cracks closed up during the cool-down period, with no permanent loss in the performance capability of the chimney.

(*e*) Immediately after the flue gas generator was shut off the surface temperature of the outer liner rose dramatically to the point where it was no longer possible to hold a hand on the liner for very long.

(*f*) Throughout the test with the flue gases flowing the surface of the outer liner remained cool to the touch.

Test No. 2—Temperature—650° C Flue Gases

The results of Test 2 are summarized in Fig. 5 which shows the surface temperatures at three points on the chimney. These three points are the hottest that occurred, and it can be seen that they did not remain below the maximum surface temperature allowed by ULC-S269-M1981.

Test No. 3—Temperature—925° C Flue Gases

The results of Test 3 are summarized in Fig. 6 which shows the same three points of the chimney and the surface temperatures at these points. Again, it

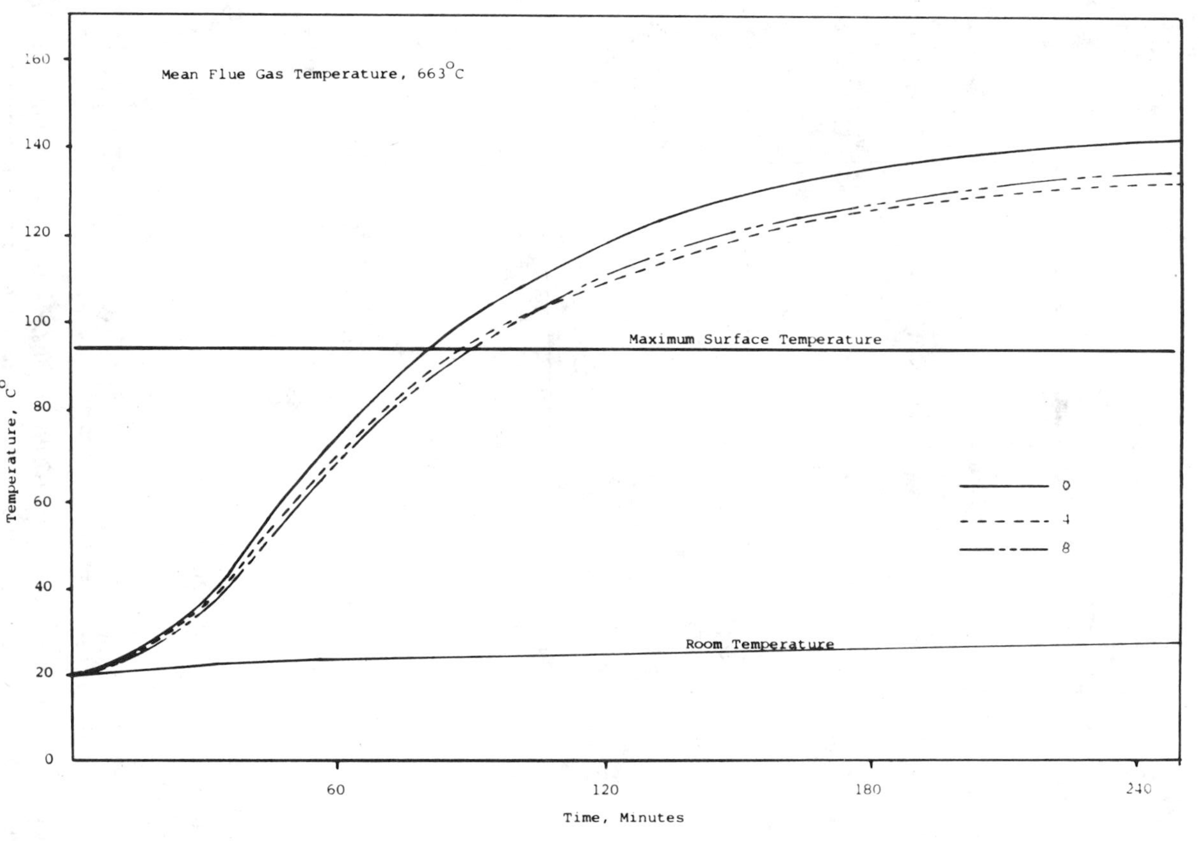

FIG. 5—*Results of Test 2.*

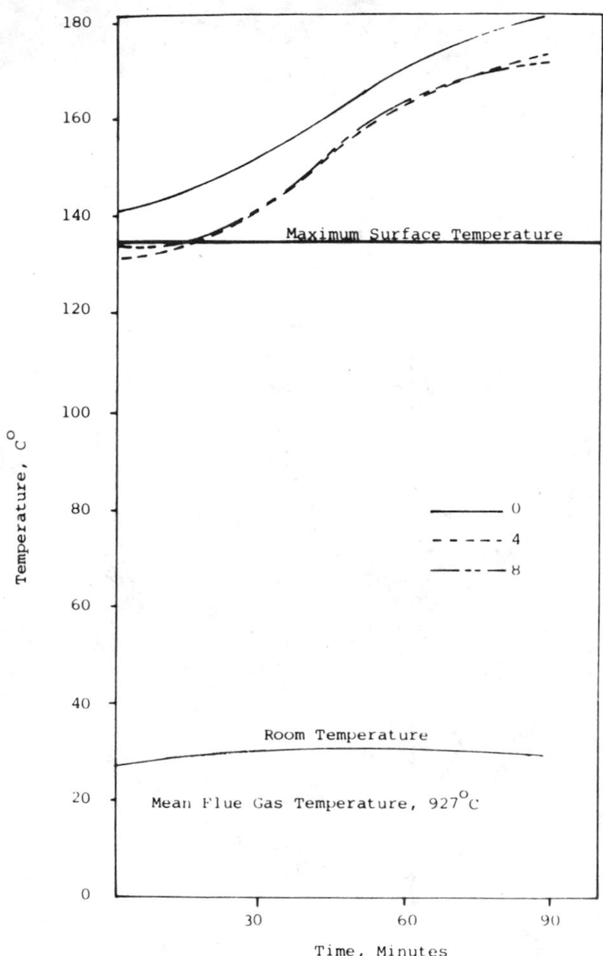

FIG. 6—*Results of Test 3.*

is apparent that the requirements for zero clearance are not being met by this chimney.

Test No. 4—Temperature—Creosote Burnout

The results of Test 4 are summarized in Figs. 7-10. Figure 7 shows the same three points as for the other tests; again it is obvious that during the creosote burnout the surface of the chimney did not meet the requirements of ULC-S629-M1981 for zero clearance.

Figures 8 through 10 give a photographic record of the deterioration of the

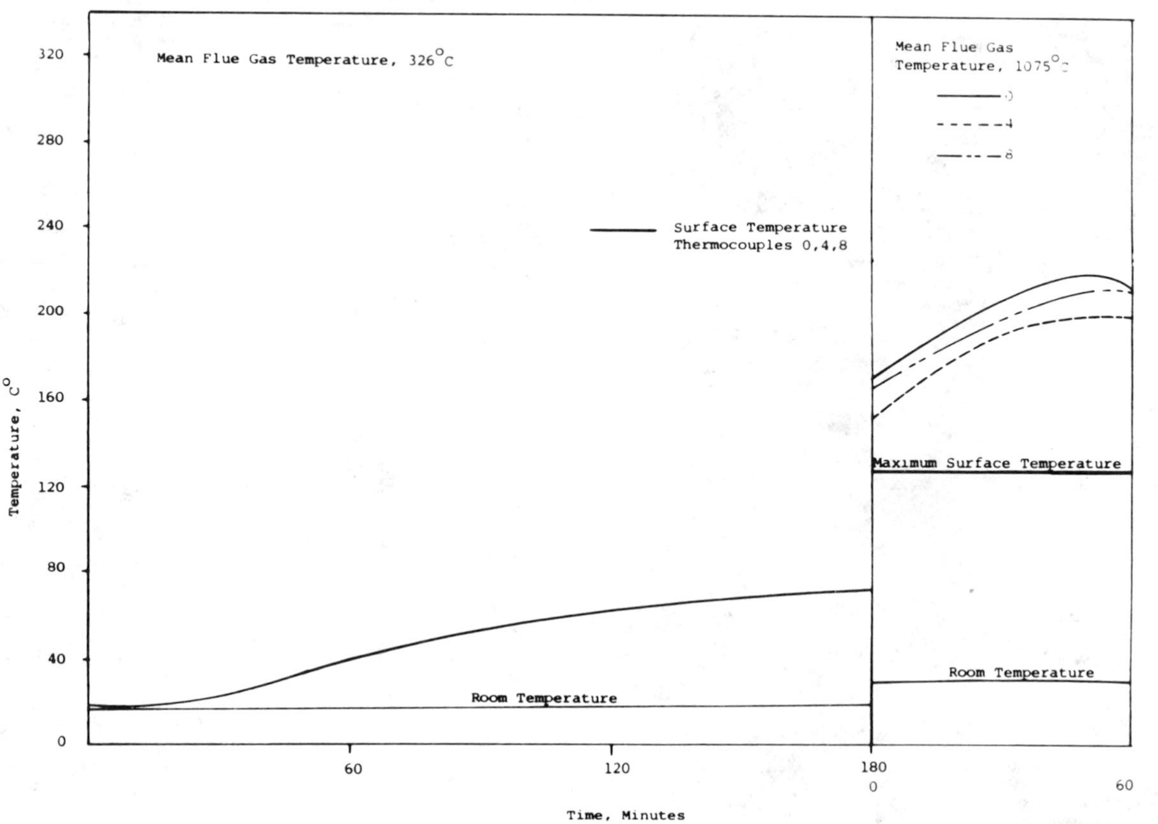

FIG. 7—*Results of Test 4.*

FIG. 8—*Joint separation.*

chimney as the test progressed and are thus self evident. Some of the cracks are only hairline, but the majority of the others were 1 mm to 7 mm in width and extended through the complete thickness of the chimney walls. During the test it was obvious when such cracks formed because of the sound which accompanied their formation. Figure 9 shows that not only did the exterior wall of the liner crack but also the interior wall.

Discussion

In accordance with engineering principles, the original unit was designed to be circular at least for the internal flue liner. However, because of the availability of standard rectangular units which were more easily moulded into the principles of the desired unit, it was decided to perform the testing program using these units.

It soon became apparent that when the liners were subjected to extreme conditions, the stress concentrations caused by the corners of their rectangular shape were too high to avoid cracking.

Testing conditions accentuated the problems caused by the connection of the generator flue pipe and the clay unit. This thimble type connection produced excessive loss of heat into the outer mantle of the lower unit, which would account for some of the external joint movement.

FIG. 9—*Internal corner crack.*

Although it is difficult to establish the exact progressive pattern of cracking under the circumstances created during the testing program, it is conceivable that the cracking started at the internal corners where temperatures were the highest, and progressively moved to the outside.

Conclusions

Tests performed at lower temperatures have shown that the proposed unit has promise to compete with the challenges provided by the metal chimney. It would appear that flue liner geometry available because of its convenience to masons may have to be sacrificed. Circular liners would have reduced the extensive cracking caused to the units tested.

In addition, further research into the possible shock resistance of clay mixtures and their modulus of elasticity would be beneficial to the eventual outcome. Such tests would have to include an in-depth study of the microstructure of clays and their behavior at elevated temperatures. Fibrous additions should be considered to achieve such goals.

With the general support of the masonry industry, a research program which will include a study of problems with conventional flue liners and ma-

FIG. 10—*External corner crack.*

sonry chimneys in general is in an advanced stage of receiving support from the federal authorities concerned.

Acknowledgments

The initial program was supported by the National Sciences and Engineering Research Council of Canada through their Project Research Applicable in Industry (PRAI) awards program. The additional program was supported by IX-L Industries of Medicine Hat, Alberta, and the Centre for Research & Development in Masonry (CRDM), Calgary, Alberta.

Special thanks goes out to Mr. David Warren of CRDM and his assistants for the execution of the testing program.

References

[*1*] Huizer, A. and Ward, M. A., "A New Clay Masonry Chimney Unit," *Proceedings of the Second North American Masonry Conference,* College Park, MD, Aug. 1982, pp. 26-2/10.

[*2*] Huizer, A. and Ward, M. A., "Performance Evaluation of an Extruded Single Unit Clay Masonry Chimney," *Proceedings of the Third Canadian Masonry Symposium '83*, Edmonton, Alberta, June 1983, pp. 24-1/16.

Ahmad A. Hamid,[1] *B. E. Abboud,*[1] *and H. G. Harris*[1]

Direct Modeling of Concrete Block Masonry Under Axial Compression

REFERENCE: Hamid, A. A., Abboud, B. E., and Harris, H. G., **"Direct Modeling of Concrete Block Masonry Under Axial Compression,"** *Masonry: Research, Application, and Problems, ASTM STP 871,* J. C. Grogan and J. T. Conway, Eds., American Society for Testing and Materials, Philadelphia, 1985, pp. 151-166.

ABSTRACT: A better understanding of the complex behavior of masonry structures is necessary to embrace the more appropriate concept of ultimate strength design. Due to the prohibitive cost of full scale testing of masonry systems, a more economical method utilizing direct modeling techniques is proposed. It is the objective of this study to evaluate the use of direct modeling of ungrouted and grouted concrete block masonry under axial compression. A total of 49 quarter-scale block prisms were tested under axial compression. Correlations between model results and available prototype tests are performed. The study includes the effects of mortar strength, grout strength, height-to-thickness ratio, number of courses, and bond type on prism compressive strength. Excellent correlations were obtained for mode of failure and moduli of elasticity. It is concluded that direct modeling is feasible and is capable of predicting the behavior of masonry. Deviations from prototype strength results were observed which are attributed to size effect of aggregate and imperfections in geometry of the model blocks.

KEY WORDS: axial compression, concrete blocks, direct modeling, grouting, masonry, mortar, models, prisms

The new concept of high-rise reinforced loadbearing masonry wall construction is an economical and structurally competitive system, particularly in the light of its added advantage of energy efficiency. Masonry codes, however, are based on working stress design because there is a lack of knowledge of the performance characteristics of masonry systems under ultimate load conditions. A better understanding of the complex behavior of masonry structures is thus necessary to embrace the more rational concept of ultimate strength design. Due to the prohibitive cost of full scale testing of masonry systems, a more economical method utilizing direct modeling techniques is proposed.

Limited studies have been performed to investigate the behavior of brick-

[1]Associate professor, research assistant, and professor, respectively, Department of Civil Engineering, Drexel University, Philadelphia, PA 19104.

work [*1,2*] and block masonry [*3–6*] using small-scale models. These studies were somewhat exploratory in nature and emphases were directed to overall performance. Reference [*5*] in particular was concerned with the elastic stage of behavior. No detailed investigation of the basic similitude requirements for direct modeling in the nonlinear range was performed which is essential for predicting the mode of failure and ultimate loads. A methodology for the direct small-scale modeling of hollow concrete block masonry has been developed by Harris and Becica [*3,4,6*] at Drexel University. Based on their limited preliminary tests, they concluded that direct modeling for concrete masonry is feasible. An extensive experimental program is currently underway at Drexel University to evaluate direct modeling of concrete masonry under different loading conditions.

Objective and Scope

It is the objective of this research to evaluate the use of direct modeling of ungrouted and grouted concrete block masonry under axial compression. Correlations between ¼ scale direct models and available prototype test data [*7,8*] are performed.

The scope of the experimental program includes studying the effects of mortar type, grout strength, number of courses, height-to-thickness ratio, and bond type on the behavior of block masonry prisms under axial compression.

Modeling Technique

The approach adopted in this study was to achieve "true" modeling; that is, to produce models which can predict the elastic as well as the inelastic behavior including failure. This necessitates obeying all similitude requirements for geometry, materials, and loading [*3*]. Using the theory of dimensional analysis [*9*], the similitude requirements can be derived for masonry (see Table 1). A more extensive discussion of the similitude requirements for masonry can be found in Refs *4*, *9*, and *10*.

Scale Factor

The geometric relationship between the model and prototype is provided by the scale factor. In this study a scale factor of four was used. The choice of this scale factor is primarily based upon two important considerations: (*a*) The Masonry Laboratory at Drexel University houses a masonry block-making machine that provides units having ¼ the nominal size of the prototype 200- by 200- by 400-mm (8- by 8- by 16-in.) blocks, and (*b*) an early study [*4*] at Drexel University indicated that the usage of scale factors greater than four would probably result in problems in the fabrication of joints. This has added signifi-

TABLE 1—*Prediction and design equations for direct modeling of masonry under static loading.*

Equations	Description
	(a) Prediction
$\epsilon_m = \epsilon_p$	strain in model is the same as in prototype
$\sigma_m = \sigma_p$	stress in model is the same as in prototype
$\delta_m = \delta_p/S_L$	deflection; where $S_L = L_p/L_m$ is the length scale
$M_m = M_p/S_L^2$	bending moment per unit length
$P_m = P_p/S_L$	compression load per unit length
	(b) Design
$t_m = t_p/S_L$	thickness
$A_m = A_p/S_L^2$	area
$E_m = E_p$	Young's modulus
$(f'_m)_m = (f'_m)_p$	compressive strength
$(f'_{tn})_m = (f'_{tn})_p$	tensile strength normal to bed joints
$(f'_{tp})_m = (f'_{tp})_p$	tensile strength parallel to bed joints
$(f'_s)_m = (f'_s)_p$	shear strength of bed joints
$\mu_m = \mu_p$	Poisson's ratio

cance in view of the importance of mortar joints in affecting the overall masonry behavior [*11*].

Materials

The constituent materials selected are scaled-down materials satisfying the basic similitude requirements for direct modeling.

Blocks

Three different ¼ scale configurations of hollow concrete blocks shown in Fig. 1 were used throughout the tests program. The ¼ scale blocks represent full scale nominal 200-mm (8-in.) blocks. The model blocks were manufactured in-house using a block-making machine and were moist-cured for 28 days. Details of the manufacturing process are included in [*10*]. The physical properties of model blocks are listed in Table 2.

Mortar

The model mortar used consisted of cementitious materials (portland cement and lime) and local Delaware River Valley masonry sand with enough water to provide adequate workability. To assure a properly scaled down joint thickness of 10 mm (⅜ in.) as is commonly used, particle sizes greater than a 600-μm (No. 30) sieve were excluded from the masonry sand. Table 3 summarizes the mix proportions and compressive strength of the three mortar mixes used in this study. For compressive strength, 51-mm (2-in.) air-cured mortar cubes were used.

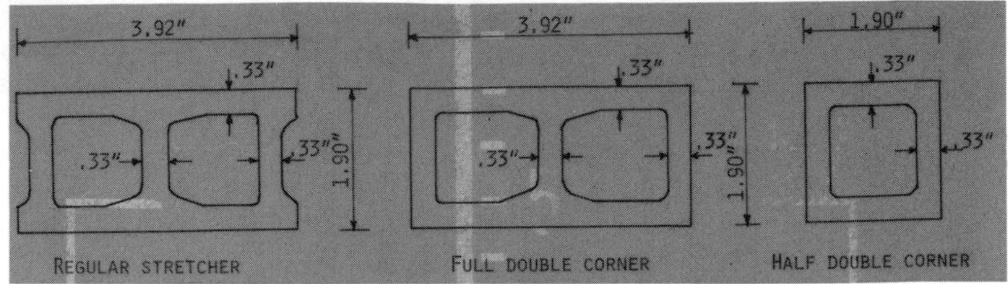

FIG. 1—*Dimensions of model blocks (25.4 mm = 1 in.).*

Grout

The sand used for model grout was the same sand used for model mortar. Three different types of grouts were used whose proportions and physical properties are listed in Table 3. Block molded prisms 25.4 by 25.4 by 50.8 mm (1 by 1 by 2 in.), which provide nearly the same surface area to volume ratio as the block cells, were used to determine compressive strength and splitting tensile strength of model grout. Specimens were air cured and tested at approximately the same average age as the corresponding prisms.

Test Specimens

Two groups of specimens (A and B) were constructed to duplicate full scale specimens [*7,8*]. Group A contains 28 three half-block-high prisms (Fig. 2) duplicating full scale prisms tested by Drysdale and Hamid [*7*] to study the effects of strength parameters such as mortar and grout strength on masonry compressive strength, f'_m. In Group B, 21 prisms with different heights and shapes (Fig. 2) were built to mirror full-scale prisms tested under axial compression by Hegemier et al [*8*]. Specimens were air cured for a minimum of 28 days before testing. This study was aimed at investigating the effects of geometric parameters such as number of courses, number of mortar joints, and bond patterns on masonry compressive strength.

TABLE 2—*Properties of model blocks.*[a]

Model Masonry Unit	Gross Area in.^{2}	Solid, %	Compressive Strength: Individual, psi	Compressive Strength: Mean, psi	Compressive Strength: COV, %	Tensile Strength: Splitting Tension, psi	Tensile Strength: Flexural Tension, psi	$\sigma_{tb}/\sqrt{\sigma_{cb}}$[b]
Regular stretcher	7.50	53	3380 3250 3810	3480	8	...	620	10.5
Full double corner	7.50	52	3730 3270 4050	3680	11	...	...	...
Half double corner	3.60	57	3220 3470 3620	3440	6	530	...	9.0

[a] 1 psi = 6.89 kN/m^2; 1 in. = 25.4 mm.

[b] σ_{cb} = block compression strength based on net area.
σ_{tb} = block tensile strength.

TABLE 3—*Properties of model mortar and grout.*[a]

	Type	Proportions (by weight) Cement:	Lime:	Sand:	Water	Compressive Strength, psi
Mortar	S1	1	0.2	3.9	1.0	1720[b]
	S2	1	0.2	4.0	1.0	1330
	N	1	0.2	4.5	1.3	860
Grout	GW	1	...	5.0	1.5	1480[c]
	GN	1	0.04	3.0	1.0	3340
	GS	1	...	2.2	0.8	4880

[a] 1 psi = 6.89 kN/m^2; 1 in. = 25.4 mm.
[b] Compressive strength of 2-in. cubes.
[c] Compressive strength of 1 by 1 by 2-in. block molded prisms.

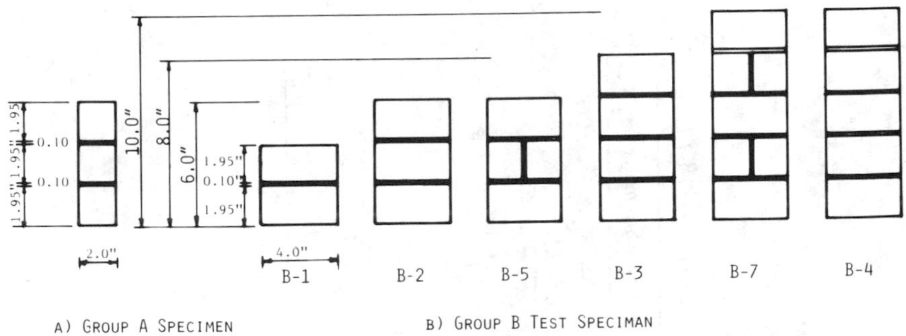

FIG. 2—*Test specimens.*

Test Procedure

Between the ages of 28 to 42 days, the prisms were capped using hydrostone and tested under axial compression. End plates were scaled down from prototype tests. Linear variable differential transducers (LVDTs) were mounted on the prisms as shown in Fig. 3 so that vertical strains could be measured. Measurements were taken at regular load increments up to approximately 90% of the failure load.

Test Results

The test results of 49 prisms are listed in Tables 4 and 5 for groups A and B, respectively. Stresses were calculated based on net area of block for ungrouted prisms and gross area for grouted prisms.

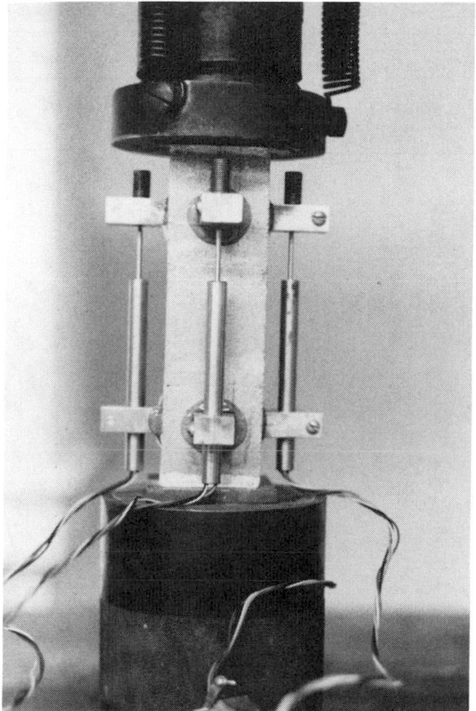

FIG. 3—*Test setup.*

Discussion of Results

Failure Mode

The failure mode for prisms having three-courses or more was a typical tensile splitting as shown in Fig. 4. For two-course prisms (Fig. 5) a shear mode of failure was observed due to end platen restraint [*11*]. Prototype results [*7,8*] indicate similar modes of failure. The similarity between model and prototype mode of failure indicates that the direct modeling technique is feasible to predict masonry behavior.

Stress-Strain Relationships

Figure 6 shows stress-strain relationships for tested model prisms along with those for prototype prisms. As can be seen, the elastic modulus of prisms with types S and N mortars were similar indicating the insignificant aspect of mortar type on assemblage modulus of elasticity. This is particularly true for grouted prisms. As the stress level increased, the effect of mortar type became signifi-

TABLE 4—*Summary of test results (Group A specimens).*[a]

Mortar Type	Grout Type	Mortar Strength[b]	Grout Strength[c]	Compressive Strength of Prisms[d] Individual, psi	Mean, psi	COV, %
S1	...	2010	...	2280 1840 2220 2210 2260	2160	8
S2	...	1730	...	2050 2170 2020	2080	4
N	...	1120	...	2040 1640 1550	1740	15
S1	GN	2010	3310	2890 2890 2530 2070 2450	2570	14
S2	GN	1730	3300	2630 2520 2730	2630	4
N	GN	1120	3300	2530 2740 2540	2600	5
S1	GS	2020	4880	3580 3040 3100	3240	9
S1	GW	2020	1480	1980 1550 2180	1900	17

[a] 1 psi = 6.89 kN/m^2.
[b] Compressive strength of air-cured 2-in. mortar cubes corrected by a factor of 1.3 to account for size effect [*4*].
[c] Unconfined compressive strength of block molded grout prism.
[d] Based on net area for ungrouted prism and gross area for grouted prisms.

cant. Prisms with type N mortar showed higher deformations compared to those for type S mortar. Similar behavior was observed for prototype prisms, particularly moduli of elasticity, indicating the feasibility of direct modeling in predicting the behavior under axial compression.

Effect of Mortar Strength

Figure 7 shows the variation of compressive strength of masonry with mortar joint strength. As can be seen, increasing mortar strength increased prism strength. The increase is more pronounced for model prisms than for the prototype, indicating the significance of mortar joints in model behavior. This is be-

TABLE 5—*Summary of test specimens (Group B specimens).*[a]

Series	h/t Ratio[b]	Number of Courses	Mortar strength,[c] psi	Grout strength, psi	Configuration	Compressive Strength Individual, psi	Compressive Strength Mean, psi	Compressive Strength COV %
B1	2	2	2450	3380	full mortar bed	2880 2990 2760	2880	4
B2	3	3	2450	3380	full mortar bed	2610 2550 2870	2680	6
B3	4	4	2450	3380	full mortar bed	2660 2610 2680	2650	2
B4	5	5	2450	3380	full mortar bed	2610 2250 2680	2510	9
B5	3	3	2450	3380	face-shell mortar bed	2920 2750 2650	2770	5
B6	2	3	2450	3380	cut, full mortar bed	2340 2800 2580	2570	9
B7	5	5	2450	3380	face-shell mortar bed	2440 2370 2410	2410	2

[a] 1 psi = 6.89 kN/m^2.
[b] h/t = height of prism/least lateral dimension.
[c] Compressive strength of air-cured 2-in. mortar cubes corrected by a factor of 1.3 to account for size effect [*4*].

FIG. 4—*Typical splitting failure of three-course model prisms.*

cause joint thickness is more controlled in the prototype than in the model. Due to the small joint thickness in the model (2.5 mm = 3/32 in.), any small variations in block height or in constructing the joints would have more impact on the strength. It was not the case, however, for grouted prisms, as shown in Fig. 7, where the continuity provided by the grout cores reduced the significance of mortar joints. It has to be noted that both model and prototype results [*8*] revealed similar trends as shown in Fig. 7.

Effect of Grout Strength

Model results indicate that the compressive strength of grout had an appreciable effect on prism strength. As shown in Fig. 8, the contribution of grouting is more significant for model prisms than it is for prototype prisms. It has to be noted that the geometry of the model grout cores is not identical to that of the

FIG. 5—*Typical shear failure of two-course model prisms.*

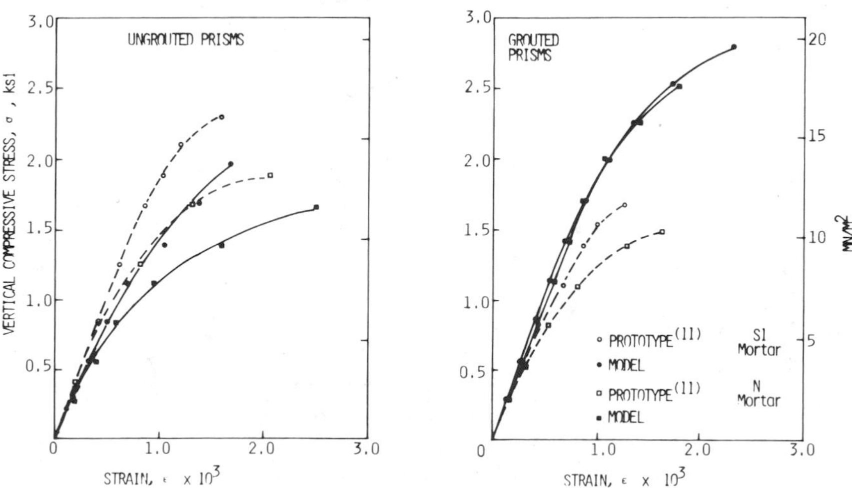

FIG. 6—*Stress-strain relationships.*

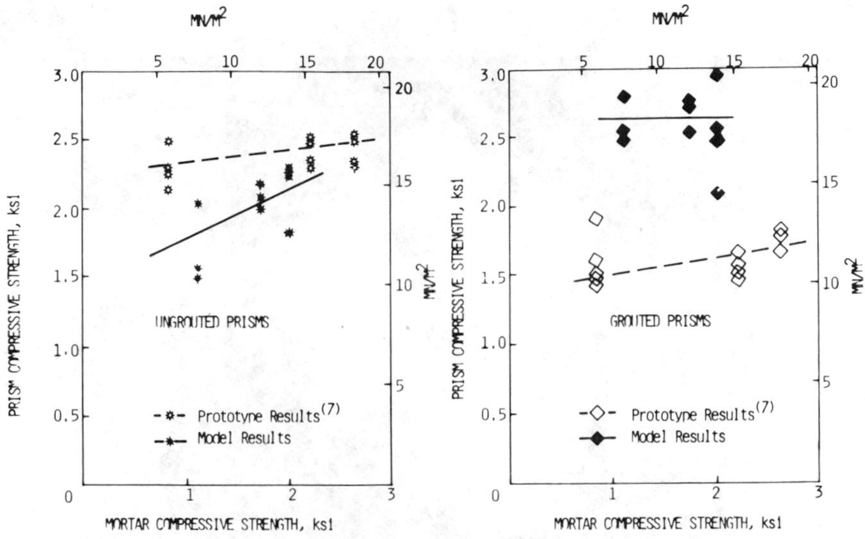

FIG. 7—*Prism compressive strength versus mortar compressive strength.*

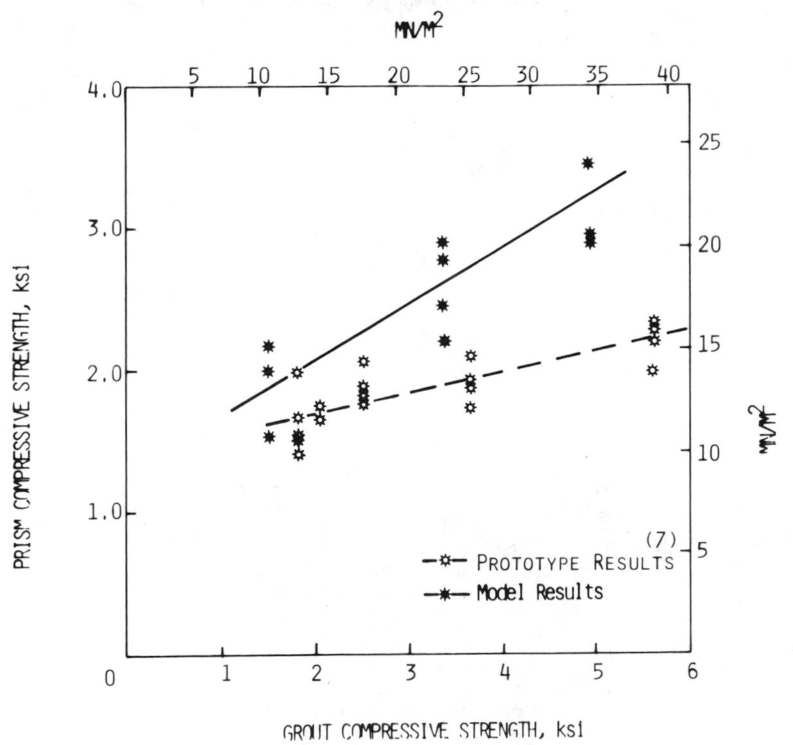

FIG. 8—*Prism compressive strength versus grout compressive strength.*

prototype. This is because full scale units have a flared shape which is not the case for model units. As a result, the critical cross section of the model grout core is larger than it should be, causing an overestimation of the contribution of grouting.

Effect of Prism Geometry

The geometry of tested prisms affected the mode of failure. For two-course prisms, shear failure (Fig. 5) was the predominant mode, whereas it was tensile splitting (Fig. 4) for prisms having three or more courses. This observation agrees with that for prototype prisms [*7,8*]. The change of failure mode is attributed to the degree of end platen restraint. For prisms of higher *h/t* (height to thickness ratio) the central portion is free from artificial confining stresses, thereby allowing splitting failure to take place [*12*].

Figure 9 shows the effect of *h/t* ratio on prism compressive strength. As can be seen, increasing the *h/t* ratio decreased prism compressive strength. The lower the *h/t* ratio the higher the confining stresses are. These stresses artificially increase compressive strength [*12*]. It has to be noted that model results agree with prototype results [*8*] in predicting the effect of prism geometry on

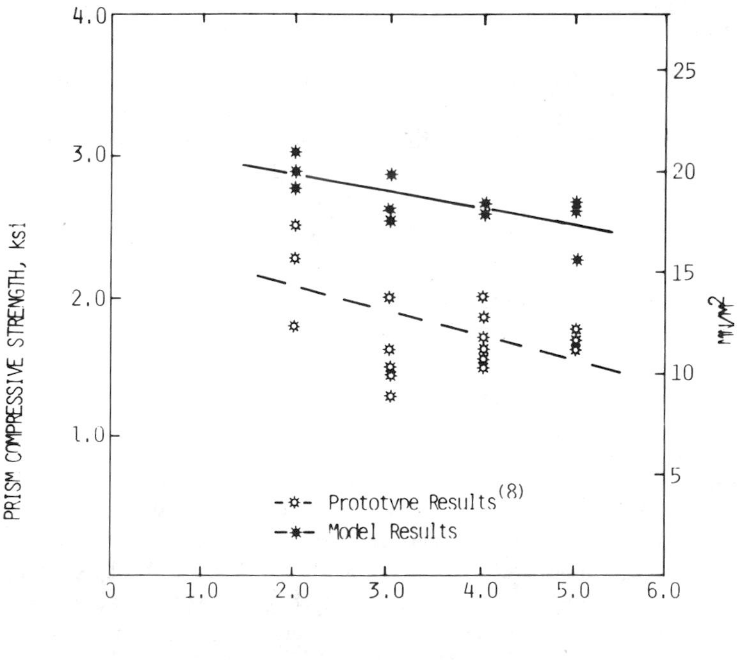

FIG. 9—*Effect of* h/t *ratio on prism compressive strength.*

compressive strength. Again, model prisms yielded higher strength values compared to prototype prisms, which is similar to what was observed previously with half-block prisms (Group A).

Comparing the results of Groups B2 and B6, Table 5 indicates the effect of number of courses on the compressive strength of prisms. The results showed that the compressive strength of prisms with $h/t = 3$ (three courses) was approximately the same as that for prisms with $h/t = 2$ (three courses with half blocks top and bottom). This indicates that prism compressive strength is a function of the number of courses and not of the h/t ratio. This conclusion agrees with Hegemier et al [*8*] prototype test results.

The effect of bond type on compressive strength can be studied by comparing the results of Groups B2 and B4 with Groups B5 and B7, respectively (Table 5). The results indicate that there is no significant effect of bond type on compressive strength of masonry prisms. This observation agrees with Drysdale and Hamid [*7*] and Hegemier et al [*8*] prototype test results.

Correlations Between Model and Prototype Results

As discussed previously, model results of the current study have compared favorably with prototype test results [*7,8*] having similar constituent material properties (see Figs. 6 through 9). Model test results revealed the following behavior which is similar to that observed for prototype masonry:

1. Failure mode changes from splitting to shear mode as h/t approaches 2.0.
2. Mortar joints have a significant effect on compressive strength of ungrouted masonry assemblages. This effect becomes insignificant for grouted masonry because of the continuity provided by the grout cores.
3. Compressive strength of masonry increases as the mortar and grout strengths increase. The increase, however, is not directly proportional.
4. Compressive strength of masonry prisms decreases as h/t ratio increases. The effect of h/t ratio is not, however, substantial.
5. Bond type (running versus stack) has no significant effect on compressive strength of masonry prisms.
6. Prism compressive strength is a function of number of courses in the prism and not of the h/t ratio.

Compared to prototype test results, model ungrouted prisms yielded lower values for compressive strength. This could be attributed to the higher sensitivity of the model prisms to variation of mortar joint thickness (2.5 mm = 3/32 in.) because of variation of block height. On the other hand, model grouted prisms provided higher values of compressive strengths than those obtained from similar prototype prisms. The discrepancy is mainly attributed to the difference in block geometry between model and prototype. Model blocks have a constant cross section with no tapering of face shells and webs. The flared shape of prototype blocks provides grout cores with variable cross sections which causes stress concentrations under applied load, and consequently a reduction in ultimate capacity [*12*].

The higher model results could also be attributed to the higher tensile to compressive strength ratio of the model units compared to that of prototype units. Because failure is caused by splitting of blocks under a combined tension-compression state of stress, an increase of tensile to compressive strength ratio leads to an increase of assemblage compressive strength [*12*]. Note that model units have oversized webs compared to prototype units (about 30% larger). This imperfection in the model units could also contribute to increase the compressive strength results of the model prisms [*4,6*].

It has been reported [*11*] that aggregate size has an effect on the behavior of concrete under combined loading and its response to confining stresses. Mortar and grout in masonry prisms under high stress levels are under triaxial compression [*12*]. It is therefore expected that behavior of model mortar joints and model grout cores is different from that of prototype materials having larger aggregate size. This difference in behavior may contribute to the deviation observed between model and prototype results.

The test results of model prisms revealed coefficients of variation similar to those obtained with prototype results. Similarity of the variability in test results indicates that the direct modeling technique is a reliable technique.

Conclusions

The present study concludes that direct modeling is feasible and capable of predicting the complex behavior of block masonry under axial compression. The ¼ scale masonry units do not resemble the exact geometry of prototype units, causing deviation in the compressive strength results from prototype results. Also, the effect of the aggregate size on the behavior of the constituent materials under multiaxial stresses could be significant in direct modeling. The effects of block geometry and aggregate size on model results are currently under investigation at Drexel University.

It is the authors' opinion that with a more refined scaling of the different geometric characteristics of the model units and a proper assessment of the effect of aggregate size a better correlation between model and prototype masonry can be achieved.

Acknowledgment

The work presented is a part of a more comprehensive test program conducted in the Department of Civil Engineering, Drexel University to study the use of direct modeling in unreinforced and reinforced masonry structures under different loading conditions. This program is supported in part through funds provided by Drexel University under its Faculty Development Mini-Grant Program. The financial support of the Graduate School at Drexel University is gratefully acknowledged.

References

[*1*] Sinha, B., Maurenbrecher, A., and Hendry, A., "Model and Full Scale Tests on Five Story Cross-Wall Structures Under Lateral Loading," *Proceedings of the Second International Block Masonry Conference*, Stoke-on Trent, England, April 1970, pp. 201-208.

[*2*] Suter, G. and Keller, H., "Reinforced Brickwork Lintel Shear Study Utilizing Small Scale Bricks," *Proceedings of the North American Masonry Conference*, Boulder, CO, Aug. 1978, pp. 24.1-24.17.

[*3*] Harris, H. G. and Becica, I. J., "Direct Small Scale Modeling of Concrete Masonry," *Proceedings EMD-ASCE, Advances in Civil Engineering Through Engineering Mechanics*, May 23-25, 1977, pp. 101-104.

[*4*] Becica, I. J. and Harris, H. G., "Evaluation of Techniques in the Direct Modeling of Concrete Masonry Structures," Structural Models Laboratory Report No. M77-1, Department of Civil Engineering, Drexel University, Philadelphia, PA, June 1977, p. 96.

[*5*] Koerner, R. M. and Rosenfarb, J., "Assessment of Fatigue Behavior of Building Foundations as Determined by Scale Models in the Laboratory," Report prepared for the United States Department of Interior, Bureau of Mines, Department of Civil Engineering, Drexel University, Philadelphia, PA, May 1978, p. 108.

[*6*] Harris, H. G. and Becica, I. J., "Behavior of Concrete Masonry Structures and Joint Details Using Small Scale Direct Models," *Proceedings of the North American Masonry Conference*, Boulder, CO, August 1978, pp. 10-1-10-18.

[*7*] Drysdale, R. and Hamid, A., "Behavior of Concrete Block Masonry Under Axial Compression," *American Concrete Institute Journal, Proceedings*, Vol. 76, No. 6, June 1979, pp. 707-772.

[*8*] Hegemier, G., Krishnamoorthy, G., Nunn, R., and Moortly, T., "Prism Tests for the Compression Strength of Concrete Masonry," *Proceedings of the North American Masonry Conference*, Boulder, CO, Aug. 1978, pp. 18.1-18.17.

[*9*] Sabnis, G. M., Harris, H. G., White, R. N., and Mirza, M. S., *Structural Modeling and Experimental Techniques*, Prentice-Hall, Inc., Englewood Cliffs, NJ, 1983, p. 585.

[*10*] Hamid, A. and Abboud, B., "Direct Modeling of Ungrouted and Grouted Block Masonry Under In-Plane Loading," Report No. MS84-1, Department of Civil Engineering, Drexel University, Philadelphia, PA, 1984, p. 98.

[*11*] Hamid, A., "Behavior Characteristics of Concrete Masonry," Ph.D. thesis, McMaster University, Hamilton, Ontario, Canada, 1978, p. 445.

[*12*] Hamid, A. and Drysdale, R., "Suggested Failure Criteria for Grouted Concrete Masonry Under Axial Compression," *American Concrete Institute Journal, Proceedings*, Vol. 76, No. 10, Oct. 1979, pp. 1047-1061.

Hong E. Wong[1] *and Robert G. Drysdale*[1]

Compression Characteristics of Concrete Block Masonry Prisms

REFERENCE: Wong, H. E. and Drysdale, R. G., **"Compression Characteristics of Concrete Block Masonry Prisms,"** *Masonry: Research, Application, and Problems, ASTM STP 871*, J. C. Grogan and J. T. Conway, Eds., American Society for Testing and Materials, Philadelphia, 1985, pp. 167–177.

ABSTRACT: This paper reports the results of part of a test series to establish stress-strain relationships for hollow, solid, and grout filled concrete block masonry. Included as part of the discussion of the interpretation of previous test programs is a comparison of the influence of the use of prisms with heights of two, three, four, and five blocks. Using the four-block-high prism as the standard, the average moduli of elasticity and the shape of the stress-strain relationship to failure are determined for the hollow, solid, and grouted cases for compression normal to the bed joint and parallel to the bed joint. Strengths for compression normal to the head joint are approximately 25% lower than for compression normal to the bed joint. In addition, strengths for grouted and solid prisms are approximately 35% lower than for hollow blocks. The moduli of elasticity also vary in nearly the same proportions.

KEY WORDS: axial loads, compressive strength, concrete blocks, grouting, loads (forces), modulus of elasticity, prisms, strain, stress

Previously published papers [*1–4*] have demonstrated significant differences in behavior for different combinations of masonry material. Specifically, grout-filled hollow blockwork had significantly lower strengths than similar ungrouted hollow blockwork. Also, the failure mechanisms were not the same as for solid masonry. The most important aspect of these results is that typical North American design provisions [*5,6*] do not differentiate in any substantial way between the above mentioned forms of block masonry. In addition, no account is taken of the direction of the compressive stress although implicitly, by providing only one value, design codes specify uniform properties for all directions of compression force.

It is apparent that where the cross webs of blocks in successive courses do

[1]Master of engineering student and professor, respectively, Department of Civil Engineering and Engineering Mechanics, McMaster University, Hamilton, Ont., Canada L8S 4L7.

not align vertically or are not joined by mortar, the compressive capacity of the masonry assemblage will be less than where force can be transmitted between cross webs at the mortar joints [*7*]. In addition, it has been shown that the incompatibility of stress-strain relationships for grout and block lead to a combined capacity which is less than the sum of the capacities of the individual components [*2,3,8*]. Therefore, differences in strength for loading normal to the bed joint are generally understood if not thoroughly explained.

For compression normal to the head joint, there is greater nonuniformity as a result of the existence of the blocks' cross webs perpendicular to the direction of the compression force. For ungrouted prisms it is likely that these cross webs provide bracing for the face shells of the block. However, it seems unlikely that they could contribute directly to load transmission. Where the cells of the blocks are grouted, these webs intercept the grout and therefore will create complex states of stress because of the changes in cross section. Although some beam tests have been reported [*9*], there does not seem to have been much research done where compression normal to the head joint is the controlling failure condition.

This paper contains the results of prism tests to document the stress-strain behavior for the above-mentioned material combinations and directions of compressive force.

Experimental Program

General Description

For convenience of testing it was decided to use prisms with a one block length cut from walls constructed in running bond. Although other researchers [*8,10,11,12*] have previously investigated the influence of prism height on capacity, it was decided to repeat this by testing prisms which were two, three, four, and five courses high for both hollow and grout-filled prisms loaded normal to the bed joint. However, for the remainder of the tests, prism heights equivalent to four courses were used based on previous tests [*8,10,13*]. These included hollow and grouted block loaded normal to the head joints and solid block loaded normal to the bed joint.

Materials

Mortor—Type S mortar was used in all prisms. The mix proportion of portland cement to lime to sand was 1:0.21:4.24 by weight or 1:½:3⅓ by volume. A constant amount of water was used giving an average flow of 122%. At the time of testing, the average age of the mortar was 17 months. The mortar cubes had been stored in dry laboratory conditions with temperature averaging approximately 21°C and relative humidity varying between 20% and 78% from winter to summer. From the three 5-cm (2-in.) mortar cubes made for each batch, 54 cubes were tested. The average compressive strength was 18.8 MPa with a coefficient of variation (*COV*) of 14.2%. Each batch was about 43.5 kg.

Grout—A medium strength grout similar to that used in previous tests [*4*] was used in all grouted prisms. The mix proportion of portland cement to lime to sand was 1:0.044:3.55 by weight or 1:0.1:3.0 by volume. This produced an average slump of 280 mm. The water to cement ratio was 0.70 by weight. At the time of testing the prisms, the compressive strength from 150 mm diameter nonabsorbent cylinder molds was 21.8 MPa. However, the more representative compressive strengths from block molded control prisms were 34.0 MPa for the standard 75 by 75 by 150 mm prisms and 30.4 MPa for 120 by 120 by 390 mm prisms. The latter prisms have a surface area to volume ratio which is very close to the conditions for grout in the cells of 190 mm hollow blocks.

Blocks—Two types of blocks were used. All were manufactured using autoclave curing. The hollow unit was a 190 mm two-cell stretcher unit. Evaluation of the minimum section at the bottom of the block including the frogged ends and the pear-shaped cells gave a net area of approximately 51% of the gross area. For running bond, the minimum contact area of the 32-mm-thick face shells was 34% of the gross area. The strength of the block was very dependent on the test conditions. For a full block hard capped with gypsum cement (Hydrostone) the strength was 19.2 MPa based on the minimum net area. However, when half blocks were tested with only 32 mm face shell hard capping, the average strength based on the minimum face shell area was 23.6 MPa. This will be discussed in more detail later.

The other unit used was a solid 190 mm block. Tests of hard-capped half blocks gave a compressive strength of 15.6 MPa based on the solid area. The full block was actually only 97.6% solid owing to a frogged end at one end of the block.

Prism Fabrication

The test specimens were made by an experienced mason over a period of three weeks. The consistency of the mortar was controlled by proportioning by weight. The amount of mortar per batch was determined by the mason's rate of work to avoid retempering. Each batch was sufficient for approximately four 4-high by 2-block-long specimens. The blocks were laid in running bond using splitter units for half blocks at the ends of alternate courses. For the three-high and five-high prisms, head joints were positioned to be at the center of the middle course of the final one-block-long prisms. All mortar joints were tooled with a cylindrical jointer. The grouting of randomly selected specimens was completed approximately two weeks after initial fabrication.

The original specimens were cut in half to provide two prisms approximately two months after fabrication. The prisms were then stored in dry laboratory conditions for an average of 17 months until they were tested.

Test Procedure

Figure 1 shows a schematic of the test setup. Steel plates 75 mm thick were used to transfer the loads to the prisms. The steel plates were placed on the ends of the prisms using approximately 3 mm thickness of gypsum cement (Hydrostone). Face shell capping was used for the hollow prisms and full capping was used for the grouted and solid units. It is estimated that the likely average maximum error in load alignment parallel to the face shells was approximately 4 mm, including the influence of crookedness of the prisms.

Axial compression was applied using a 2500-kN-capacity test machine. A loading rate of approximately 1 mm/min. was used but loading was stopped at approximately twelve equal load steps to take strain readings. The total time for testing was approximately 30 min. Strain readings were taken using both mechanical and electrical measuring instruments. Readings were taken over the central position of the specimens and included both 20 mm and 200 mm gage lengths spanning a joint. Four gage lengths for each type were averaged for each specimen and were located at the quarter points on each side of a specimen.

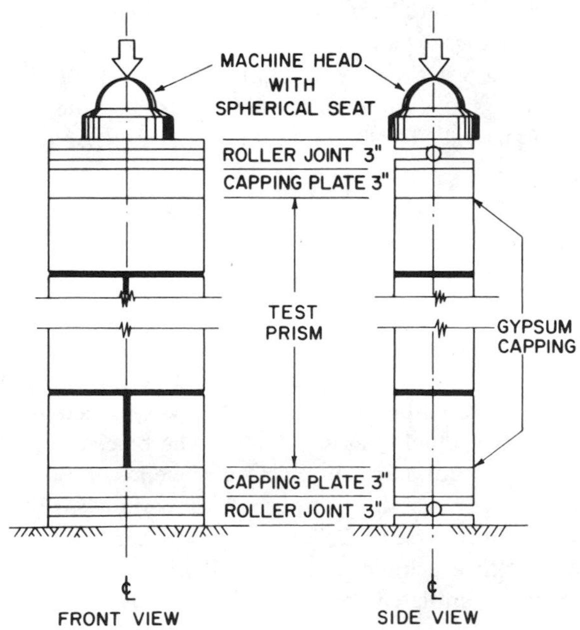

FIG. 1—*Schematic of test setup.*

Discussion of Test Results

Failure Modes

The failures of the hollow block prisms loaded normal to the bed joint were always initial splitting of the cross webs followed by instability and collapse of the face shells at higher levels of load. This failure of the face shells sometimes included a type of shearing through of the face shells, particularly in the blocks at the ends of the prisms. Therefore, for the two-course prisms, this conical failure appearance tended to predominate at least on one face of the prisms.

Figure 2 contains photographs showing failures of grouted prisms. For loading normal to the bed joint (Fig. 2*b*), the failure pattern was quite similar to that for hollow prisms. However, the presence of the grout limited the degree of shearing or splitting through the face shells. Conversely, it was not uncommon to have face shells from two or more courses included in the failure zone. No damage was apparent in the grout [*14*]. For the prisms made using solid units, the initial vertical crack followed a plank through the head joints. Final failure occurred when these cracks extended along the bed joints and through other parts of the blocks.

For tests with compression normal to the head joint, the initial cracking occurred along the bed joint between the two courses. For the hollow prisms the final failure followed closely with buckling of the face shells. For the grouted prisms the final failure was marked by growth of vertical crack between the face shells and the grout. This crack crossed the face shells in a shearing mode at failure as shown in Fig. 2(*a*). This final failure was fairly obviously affected by the presence of the slot between the center webs of the splitter blocks which were used at the ends of one course. However, the regions over which strains were measured were not affected.

Compressive Strengths and Stress-Strain Relationships

Table 1 contains a summary of the prism test results. From Series 1 and 2 for hollow and grouted prisms respectively, it is apparent that a four course prism height is desirable to minimize the influence of end restraint while not having any significant slenderness effect. The three, four, and five-course prisms had nearly constant capacities. The strengths for ungrouted prisms had nearly constant capacities. The strengths for ungrouted prisms were based on minimum face shell area whereas those for grouted prisms were based on gross area.

Using the four block-high prism as the standard, it can be shown that the strength of grouted prisms is approximately 35 percent less than for hollow prisms. This is applicable for compression normal to both the bed and head joints. For both grouted and hollow prisms, the strength for compression normal to the head joints was approximately 25 percent lower than for compres-

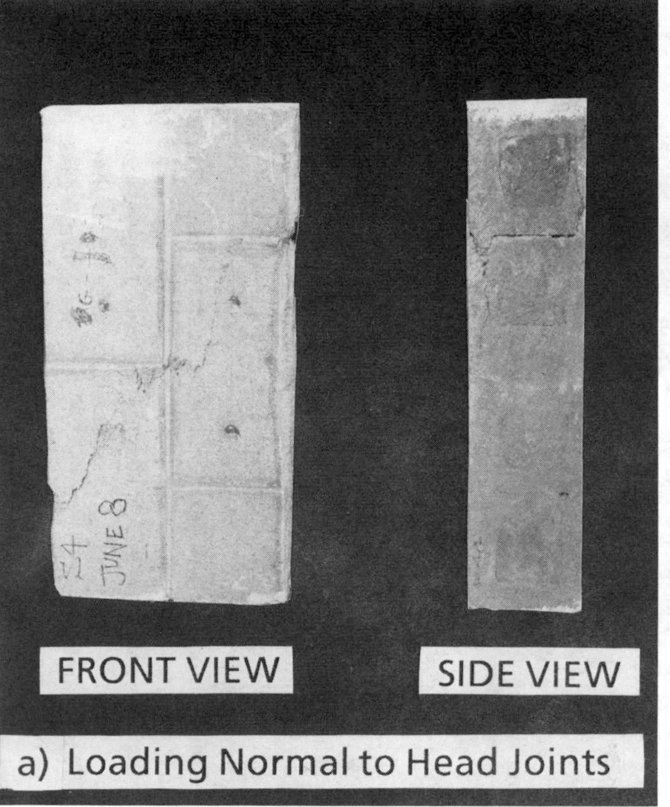

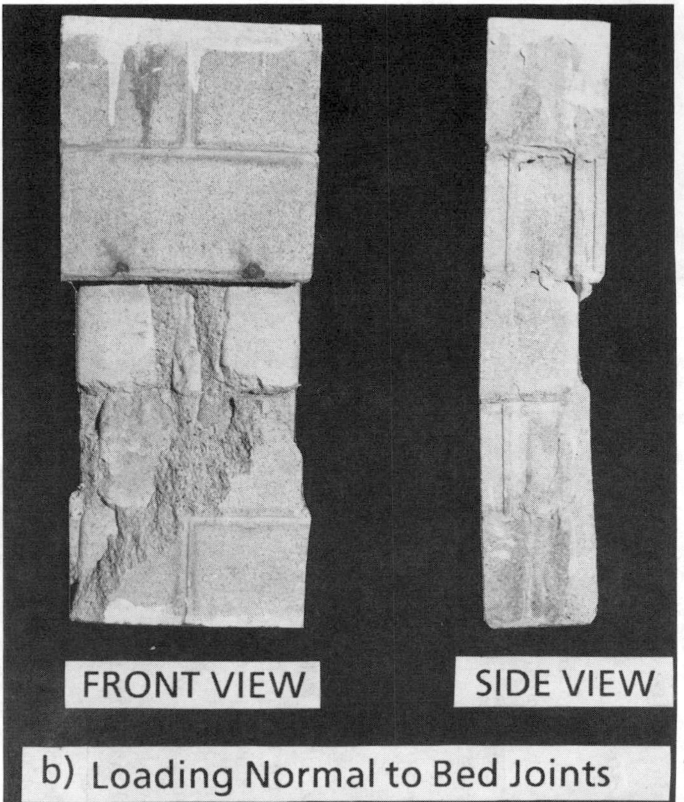

FIG. 2—*Failure modes for grouted prisms.*

TABLE 1—*Prism compressive strengths.*

Prism Series	Block Type	Direction of Load	Grouted (Y/N)	Number of Courses	Number of Tests	Mean f'_m, MPa	*COV*, Percent
1	hollow	normal to bed joint	N	2	3	24.8	10.4
				3	3	21.9	6.0
				4	3	22.5	5.7
				5	3	22.4	4.8
2	hollow	normal to bed joint	Y	2	3	18.8	11.6
				3	3	14.9	9.5
				4	3	14.5	0.6
				5	3	13.0	3.5
3	solid	normal to bed joint	N	4	4	14.7	5.8
4	hollow	parallel to bed joint	N	4	3	17.3	7.3
5	hollow	parallel to bed joint	Y	4	3	10.6	15.6

sion normal to the bed joint. The secant moduli of elasticity at a 5 MPa stress also varied in nearly the same proportions as the strengths and were generally in the vicinity of 1000 times the compressive strength.

Figures 3 and 4 contain plots of the average stress-strain relationships for the prism tests. Each point represents the average of the prisms using four readings on each prism. To show the range of results, the bands within the dashed lines encompass all the individual prism test data. The shapes of the stress-strain curves are similar for similar prisms and, if based on stress as a ratio of strength, would be nearly coincident.

Figure 5 contains a plot of the stress-strain curve for the block-molded grout control prisms. Also, the average stress-strain curve for full hard-capped blocks is shown. In addition, the average strain over a 20 mm gage length including a bed joint is shown for loading normal to the bed joint for grouted prisms. (The bands within the dashed lines encompass the ranges of results from individual tests.) It is apparent that the presence of the mortar joint and the effect of the nonuniform cross-sectional geometry lead to a much lower stiffness for this bed joint region of the assemblage. By comparison with Fig. 3(*a*) it is also obvious that the 20 mm bed joint region has a much lower modulus of elasticity than was observed for an average 200 mm length for grouted prisms.

Comparison of the stress-strain relationship for the block in Fig. 5 with that for the hollow prism in Fig. 3(*a*) can lead to some misunderstanding. The results show that the prism is both stronger and stiffer than the block. For proper interpretation, however, it must be remembered that stresses for the hollow prisms were based on minimum face shell areas. This ignores the fact that the integral webs of the blocks must participate in carrying the load over

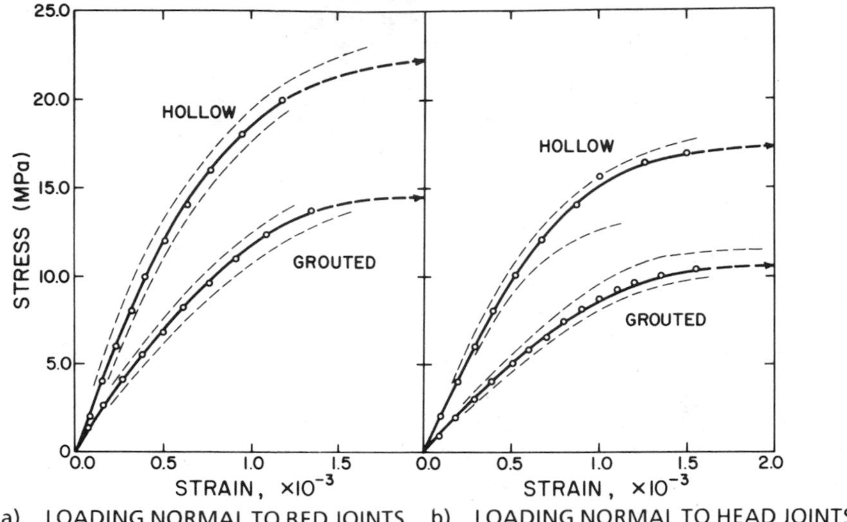

FIG. 3—*Stress-strain relationships for hollow and grouted prisms.*

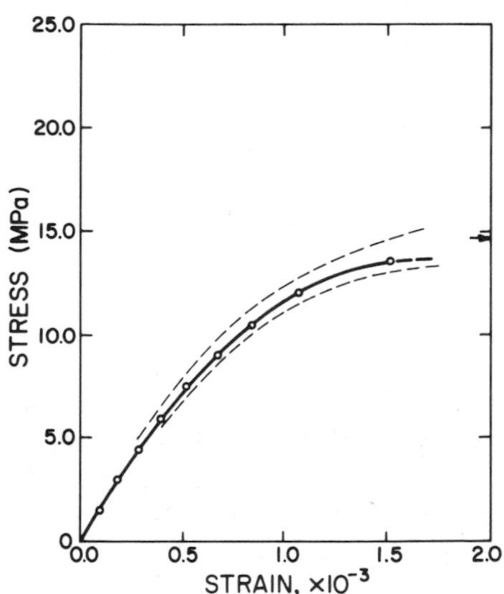

FIG. 4—*Stress-strain relationship for solid prism.*

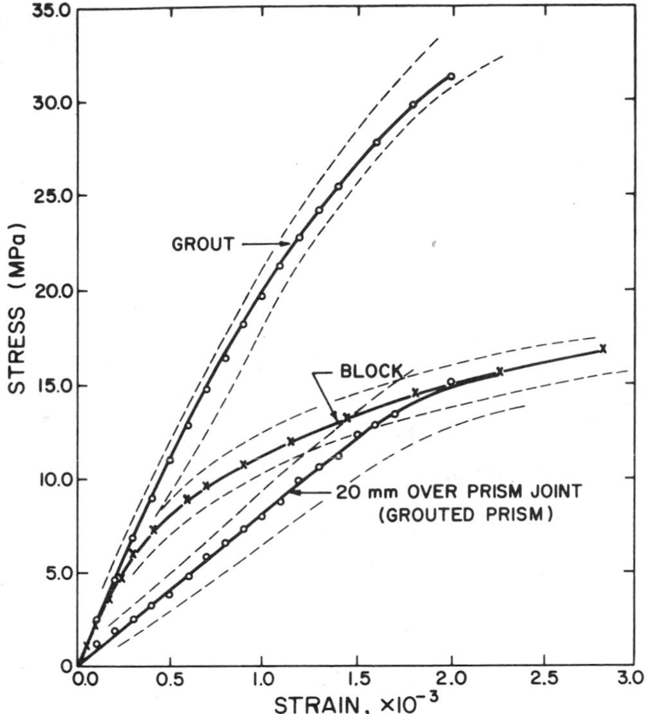

FIG. 5—*Stress-strain relationships for block, grout, and prism joint region.*

at least part of the height of the block [*15,16*]. Use the minimum area of the block amounts to an increase of 50% over the face shell area of the block. It is only the mortar which is limited to the face shell areas.

Conclusions and Recommendations

1. These results confirm that the strength and stress-strain characteristics of prisms made with similar units differ markedly depending on the direction of compression and whether the prisms are hollow, solid, or grouted. These differences are sufficiently large to warrant their separate treatment in design standards.

2. The failure mechanisms for one-block-wide prisms seem to be similar to those for full walls [*17*]. However, correlation of wall strengths and prism strengths can only be accomplished if similar construction is used in both cases.

3. The use of regular two-core stretcher units for grouted masonry is not very efficient where compressive strength is the governing criterion. Therefore, it is suggested that availability of units which result in aligned webs would be beneficial.

4. It is suggested that four-course prism heights be used as a standard for block research. This assures equal influence of bond pattern.

5. There is considerable confusion regarding both the definition and use of block compressive strength. It should be obvious that compression test of full blocks using full capping do not represent the actual behavior in a wall with face shell bedding. On the other hand, tests of blocks with only face shell bedding do not correctly measure the material characteristics.

Acknowledgments

This research was carried out at McMaster University and was funded by Operating Grants from the Natural Sciences and Engineering Research Council of Canada and the Masonry Research Foundation of Canada. The authors appreciate the contribution of the mason's time made available through the Ontario Masonry Contractor's Association and the Ontario Masonry Promotion Fund, and we thank the Ontario Concrete Block Association for the donation of the blocks.

References

[*1*] Hamid, A. A., Drysdale, R. G., and Heidebrecht, A. C., "Effect of Grouting on the Strength Characteristics of Concrete Block Masonry," *Proceedings of the North American Masonry Conference*, Boulder, CO, Aug. 1978, pp. 11-1, 11-17.

[*2*] Drysdale, R. G. and Hamid, A. A., "Behavior of Concrete Block Masonry Under Axial Compression," *American Concrete Institute Journal, Proceedings*, Vol. 76, June 1979, pp. 707-721.

[*3*] Hamid, A. A. and Drysdale, R. G., "Suggested Failure Criteria for Grouted Concrete Masonry Under Axial Compression," *American Concrete Institute Journal, Proceedings*, Vol. 76, No. 10, Oct. 1979, pp. 1047-1061.

[*4*] Drysdale, R. G. and Hamid, A. A., "Capacity of Block Masonry Under Eccentric Compression Loading," *American Concrete Institute Journal*, Proc. Vol. 80, No. 2, March-April, 1983, pp. 102-108.

[*5*] ACI Committee 531, "Building Code Requirements for Concrete Masonry Structures," ACI 531-79, American Concrete Institute, Detroit, MI. 1978, pp. 1-20.

[*6*] Canadian Standards Association, "Masonry Design and Construction for Buildings," (CSA S304-M78), Rexdale, 1978.

[*7*] Drysdale, R. G. and Hamid, A. A., "Influence of the Characteristics of the Units on the Strength of Block Masonry," *Proceedings of the Second North American Masonry Conference*, University of Maryland, College Park, MD, August 1982, Paper No. 2, 13 pp.

[*8*] Boult, B. F., "Concrete Masonry Prism Testing," *American Concrete Institute Journal*, Proc. Vol. 76, No. 4, April 1979, pp. 513-535.

[*9*] Keller, H. and Suter, G. T., "Variability of Reinforced Concrete Masonry Beam Strength in Flexure and Shear," *Proceedings of the Second North American Masonry Conference*, University of Maryland, College Park, Md., Aug. 1982, pp. 10-1, 10-14.

[*10*] Hegemier, G. A., Krishnamoorthy, G., Nunn, R. O., and Moorthy, T. V., "Prism Tests for the Compressive Strength of Concrete Masonry," *Proceedings of the North American Masonry Conference*, Boulder, CO, August 1978, paper 18, pp. 18-1, 18-17.

[*11*] Maurenbrecher, A. H. P., "Effect of Test Procedures on Compressive Strength of Masonry Prisms," *Proceedings of the Second Canadian Masonry Symposium*, Ottawa, June 1980, pp. 119-132.

[*12*] Read, J. B. and Clements, S. W., "The Strength of Concrete Block Walls—Phase II: Under

Axial Load," Technical Report, Cement and Concrete Association, London, September 1972.

[*13*] Maurenbrecher, A. H. P., "Use of the Prism Test to Determine Compressive Strength of Masonry," *Proceedings of the North American Masonry Conference*, Boulder, CO, August 1978, No. 91.

[*14*] Miller, M. E., Hegemier, G. A., and Nunn, R. O., "The Influence of Flaws, Compaction, and Admixture on the Strength and Elastic Moduli of Concrete Masonry," *Proceedings of the North American Masonry Conference*, Boulder, CO, August 1978, pp. 17-1 to 17-17.

[*15*] Maurenbrecher, A. H. P., "Compressive Strength of Eccentrically Loaded Masonry Prisms," *Proceedings of the Third Canadian Masonry Symposium*, Edmonton, Canada, June 1983, pp. 10-1 to 10-13.

[*16*] Ameny, P., Loov, R. E., and Shrive, N. G., "Prediction of Elastic Behavior of Masonry," *International Journal of Masonry Construction*, Vol. 3, No. 1, 1983, pp. 1-9.

[*17*] Read, J. B. and Clements, S. W., "The Strength of Concrete Block Walls—Phase III: Effect of Workmanship, Mortar Strength and Bond Pattern," Technical Report, Cement and Concrete Association, London, Nov. 1977.

Byron Johnson[1]

Towards a Decision-Making Strategy for Masonry Cleaning

REFERENCE: Johnson, B., **"Towards a Decision-Making Strategy for Masonry Cleaning,"** *Masonry: Research, Application, and Problems, ASTM STP 871*, J. C. Grogan and J. T. Conway, Eds., American Society for Testing and Materials, Philadelphia, 1985, pp. 178–181.

ABSTRACT: The author proposes a decision chart for the use of building owners, designers, and contractors to assist with the selection of an appropriate technique for cleaning masonry walls. The decision chart considers the nature of the masonry, the source of the soiling, and most importantly the reason for the cleaning initiative. The decision chart is presented as a suggested approach and the author requests constructive criticism.

KEY WORDS: masonry, cleaning, quality control

Cleaning of masonry has been a contentious issue for well over a century; but now as an industry of over $300 million in North America alone it has become an established component of the building industry [*1*]. There are however many differing opinions and much confusion regarding the most appropriate technique. There has been a great deal of advice from experts; most of it has been didactic and prescriptive. There has been a continual search for a universal technique, despite the fact that masonry is a highly variable material. There has been out-right condemnation of some techniques, notably sandblasting, now said to be in "disrepute," [*1*], despite the fact that for some masonry such condemned techniques may be most appropriate. There has been an obsession among professionals with esoteric techniques, despite costs.

These dogmatic approaches to the cleaning of masonry have left conscientious owners, program administrators, and contractors confused without any way of finding a way through the mass of often contradictory advice. What is needed is a method or strategy to help such persons select the most appropriate technique for the cleaning of specific masonry structure. Such a method will

[1]Senior restoration maintenance officer, Restoration Services Division, Engineering and Architecture Branch, Parks Canada, Ottawa, Canada.

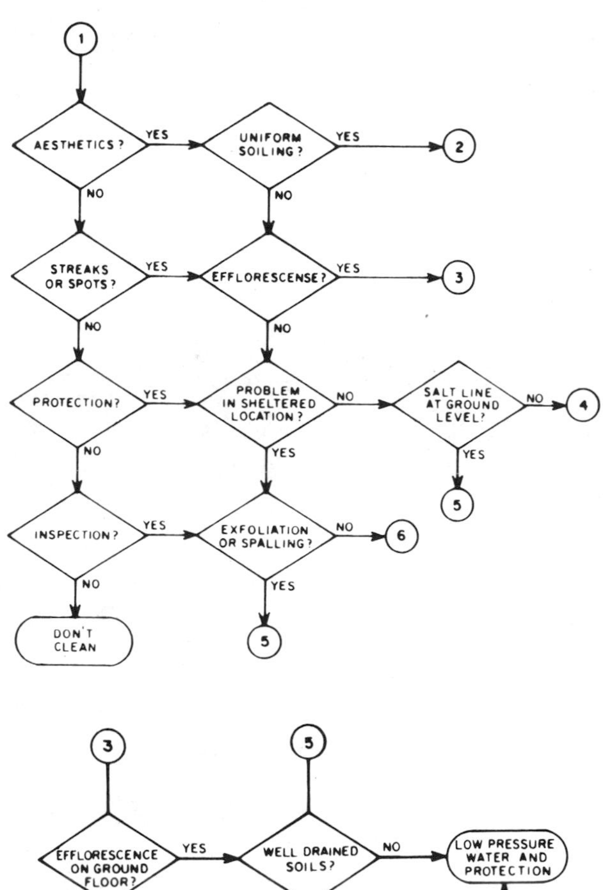

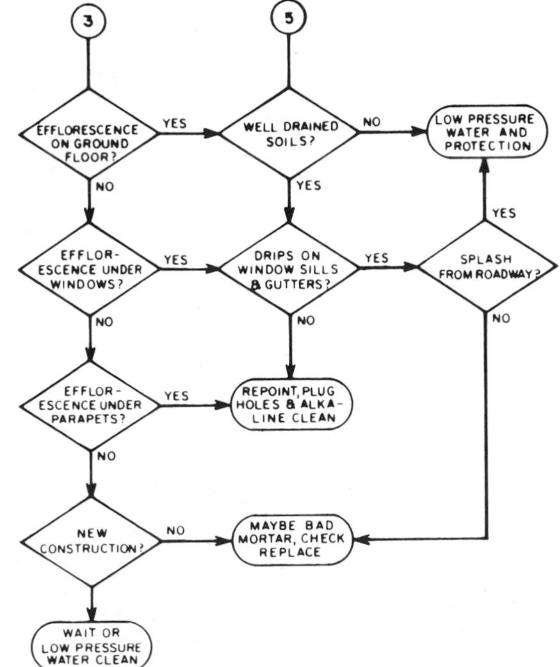

FIG. 1—Decision tree.

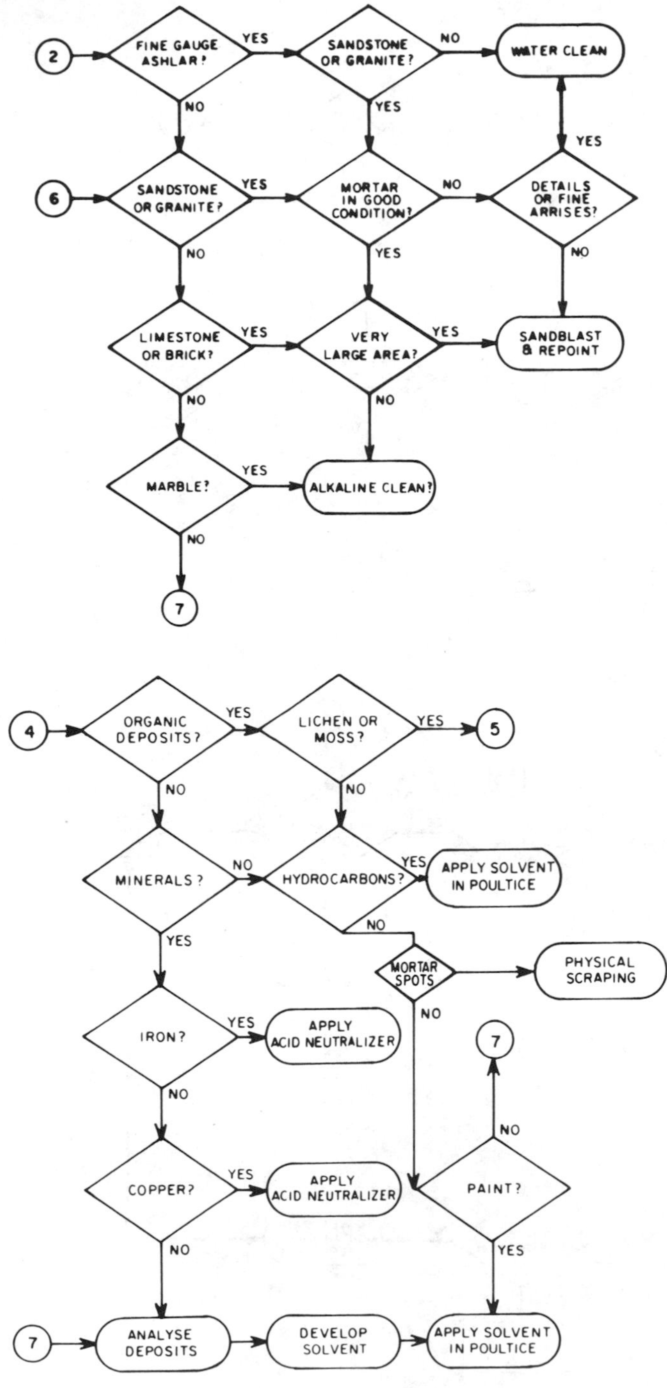

FIG. 1—*Continued.*

have to consider several attributes of the masonry and the realities of the client's needs. It is relatively easy to list the attributes of masonry that will determine the choice of cleaning; but to allow them to be used for decision-making the attributes need to be ranked by priority. The following hierarchy is presented for consideration and comment:

1. Purpose of cleaning: is the owner concerned strictly with image, is there a need to remove damaging deposits, is cleaning necessary for undertaking a thorough inspection?
2. Nature of the deposits: are they stains of rust, asphalt, and so forth, are they salts for efflorescence, are they soot or precipitated salts, are they rendering of paint?
3. Nature of the masonry units: are they hard or soft brick, are they calcareous, are they hard sandstones, are they free of iron, are there arrisses, is there details or tooling of importance?
4. Nature of masonry: is it fine gaged or coarse, is the mortar in good condition, are there details, is the masonry high (that is, is scaffolding needed)?
5. Contract realities: how large is the budget, how sensitive is the environment, what time of year is the work to be done, in what weather conditions?
6. Quality control: is the contractor reliable, what degree of supervision is possible?

Taking these attributes together it is possible to construct a decision chart that can be used for most situations by a person with the pertinent facts (Fig. 1). Some authors have recommended, quite logically, the selection of masonry cleaning based on the two attributes of the nature of the deposit and the nature of the masonry [*2*].

These decision-making methods need to be reviewed carefully, modified when necessary, and tried for some time before becoming part of a more general strategy for rehabilitation.

References

[*1*] Desson, K. and Weaver, M., "The Business of Cleaning: Preservationists mount a war on grime." *Canadian Heritage*, February-March 1983, pp. 25-26.
[*2*] Staff, "Masonry Restoration." *Construction Canada*, May 1982, Vol. 24, No. 3, pp. 9-15.

Charles H. Raths[1]

Brick Masonry Wall Nonperformance Causes

REFERENCE: Raths, C. H., "**Brick Masonry Wall Nonperformance Causes,**" *Masonry: Research, Application, and Problems, ASTM STP 871*, J. C. Grogan and J. T. Conway, Eds., American Society for Testing and Materials, Philadelphia, 1985, pp. 182–201.

ABSTRACT: The paper reviews and discusses typical nonperforming characteristics of brick masonry walls in regards to brick and mortar water penetrations, durability, and efflorescence. Factors associated with nonperformance are design, specifications, ASTM requirements for masonry, and construction workmanship. Typical nonperforming conditions are presented and methods used to evaluate brick masonry wall performance are given. Recommendations are offered to prevent nonperformance.

KEY WORDS: barrier wall, brick masonry, cavity wall, construction, curtain wall, design, durability, efflorescence, failures, investigation, leaks, permeability, water penetration, workmanship

The in-place service performance of brick masonry curtain walls in certain instances has not kept pace with the technological advancements of materials used in construction of the walls nor the structural frames to which the walls are attached. Thinner (reduced overall thickness) brick curtain walls, more sophisticated structural frames with longer spans, more complex interfacing requirements of the various curtain wall components, differential movement behavior between the brick curtain wall and the supporting structural frame, the use of new materials, and present day economics are in general the aspects associated with brick wall nonperformance.

The objectives of this paper are to review the more common nonperformance categories or causes and discuss the typical factors involved in nonperformance, the methods of evaluating nonperformance, and the prevention of nonperformance in both relatively new buildings and older buildings.

[1]Senior principal, Raths, Raths & Johnson, Inc., Willowbrook, IL.

Nonperformance Categories

Brick masonry curtain walls can be constructed in a number of ways. They can be double-wythe brick and concrete block, or a single wythe of brick. The double-wythe walls may utilize a nominal 51-mm (2-in.) cavity between the brick and the block, employ a grout-filled collar joint, or have the brick parged on its collar joint side. The single-brick wythe wall may consist of brick only or brick laterally supported by metal studs.

Regardless of the wall type, the function of an exterior brick curtain wall is to provide a permanent barrier for protecting the building interior against the outside elements. The principal nonperforming categories of walls are water penetration, durability, efflorescence, and structural behavior.

Water Penetration

Experience indicates that water penetration through masonry exterior walls is the most common cause of complaints regarding nonperformance. Water leakage manifests itself by damage and stains to exterior walls, interior side wall coverings, and ceilings adjacent to the exterior walls, water-soaked floor coverings, standing water on floors, dripping water on window sills, and the like. The penetrating water can lead to deterioration of interior materials such that replacement is required, corrosion of supporting structural steel occurs, damage to electrical and mechanical systems takes place, and the effectiveness of certain insulation types is reduced.

Depending upon the cause of the water penetration and the magnitude of it, the durability of bricks and mortar can be reduced. This results from excess water penetration which may cause the exterior brick wythe, and its mortar, to become critically saturated and vulnerable to freezing damage.

It is well established in the technical literature [*1*] that water will penetrate through a brick wythe. Brick walls constructed with materials of accepted quality (satisfying standard ASTM requirements) and acceptable workmanship quality permit water penetrations through the brick-mortar interface of the head and bed joints, as shown by the extreme Fig. 1 condition. Thus, the wall design, material selection, and quality of construction are important in determining the brick wall's resistance to water penetration. The general causes of brick wall water penetration into the building interior are

1. Head and bed joints not full, or joints with furrows or other types of voids.
2. Mortar joints of excess width and not correctly tooled.
3. Incorrect selection of the wall's materials regarding the physical properties of the brick and the mortar.
4. Mortar not compatible with the brick thereby resulting in poor bond.
5. Exterior brick walls not being either a cavity or barrier type.
6. Flashing details considering only typical conditions, and the details not interfacing with other construction materials and these other materials' space requirements creating noncontinuous flashings, Figs. 2 and 3.

FIG. 1—*Commonly observed separation gap, mortar to brick, about the brick perimeter.*

FIG. 2—*Noncontinuous metal flashing at intersecting walls. Note mortar droppings on flashing.*

FIG. 3—*Noncontinuous fabric flashing at building column.*

7. Inadequate consideration of differential movement between the exterior brick wall and the supporting structural frame relative to brick moisture expansion, brick temperature expansion and contraction, supporting frame structural deformations, temperature differentials, and applied wind or other loads.

8. Lack of construction testing and inspection by knowledgeable and experienced personnel.

Durability

Durability of brick, as discussed herein, relates to freeze-thaw cracking, spalling, and disintegration of brick. Generally, freeze-thaw problems develop when brick becomes critically water saturated as illustrated by Figs. 4, 5, and 6. A brick can be considered critically saturated when its internal moisture content is greater than the brick's 24 h cold water absorption.

Bricks become critically saturated for a variety of reasons. Among some of the more common causes are

1. Excess water penetrations through head and bed joints.
2. Inoperative or defective flashings.
3. Weep holes plugged by mortar droppings.
4. Copings, typically on parapets, which permit the entrance of excess water.
5. Window leakage and drainage.
6. Coatings applied to the exterior brick face which prevent proper breathing of the brick.
7. Inadequate venting of the brick wall's space between the wythes preventing breathing of the brick.

Glazed brick represents a special case relative to the brick's face being impervious to water and preventing breathing. In this instance, the other factors

FIG. 4—*Free-standing wall freeze-thaw deterioration of face brick.*

FIG. 5—*Freeze-thaw damage to parapet wall face brick.*

FIG. 6—*Building glazed brick freeze-thaw failure.*

contributing to critical saturation become more significant and important if critical saturation is to be avoided. Face, building, or glazed brick all suffer freeze-thaw failures basically in the same manner when they are critically saturated with water.

Investigations of brick walls undergoing freeze-thaw failures reveal that the behavior of bricks in actual walls differs from the behavior of bricks in the testing laboratory. Bricks in walls are subjected to directional freezing rather than the uniform multidirectional freezing which occurs during laboratory testing. Directional freezing results in temperature differentials across the brick creating cyclic flexural stresses. Directional freezing also is suspected to result in greater amounts of moisture in the brick at the colder exterior face, perhaps resulting in more critical saturation adjacent to the freezing face. For these reasons, and possibly because of other unknown physical behaviors, bricks can satisfy all ASTM requirements and have successfully passed ASTM Standard for Sampling and Testing Brick and Structural Clay Tile (C 67-81) freeze-thaw tests but fail in service by freezing and thawing.

Critical saturation appears to be the element in determining whether a brick in service will fail by freeze-thaw behavior. Another factor believed to be important to a brick's durability is the size and distribution of the brick's internal pores. Core holes in the brick do not seem to play any role in brick durability behavior.

Efflorescence

Efflorescence, in most instances, represents an aesthetic nonperformance category of masonry walls. As used in this paper, efflorescence refers to the white deposits which appear on the exterior face of brick walls.

Efflorescence is typically caused by water soluble salts in solution brought to the exterior face of the brick and subsequently deposited upon the brick's face by evaporation. The more common efflorescence salts, according to the technical literature [*2, 3, 4*], are aluminum, calcium, magnesium, potassium, and sodium carbonates and sulfates. The sources of the salt which generally create the efflorescence problems are the mortars and concrete masonry backup materials. The main causes of efflorescence result from

1. Brick or concrete masonry backup mortars using portland cement having high alkali contents as either sulfates or as the hydroxide.
2. Mortars employing high contents of cement.
3. Mortars which have low lime contents.
4. Concrete masonry backup units having high contents of soluble salts.
5. Walls which lack adequate resistance to water penetrations or which do not allow penetrated water to drain away.
6. Walls where the moisture enters as vapor from the building interior and accumulates within the brick as the vapor condenses.
7. Walls lacking internal ventilation, such as the cavity type, restricting the elimination of water.

An infrequent cause of efflorescence is the brick itself because of it containing water soluble salts. However, most bricks that comply with ASTM Specification for Building Brick (Solid Masonry Units Made from Clay or Shale) (C 62-81) or Specification for Facing Brick (Solid Masonry Units Made from Clay or Shale)

(C 216-81) commonly do not contribute to efflorescence but effloresce due to the aforementioned causes.

Efflorescence in the northern climates is considerably more noticeable in the late fall, winter, and early spring. This results because of the physics [*2*] involved with moisture migration and salt solubility relative to decreased temperatures. At colder brick surface temperatures, the amount of salt a solution can hold is reduced resulting in increased deposits of the salt. Also, a decrease in temperature results in further condensation of vapor within the brick and subsequent moisture migration to the colder surface where, because of decreased solubility, additional salt is deposited. A sudden temporary increase in temperature during colder months reverses the process causing the frequently observed short term reduction, or disappearance, of efflorescence.

Structural Behavior

Nonperformance pertaining to structural behavior results in cracking of brick masonry walls. Structural behavior affecting brick walls in practice reflects the integrated behavior of the brick wall with the supporting structural frame. When the behavior of the supporting frame and brick wall are not compatible, cracking results. Some of the more common causes of brick wall cracking are

1. Deflections of steel angle lintels.
2. Deflections of structural beams which support the brick.
3. Movement between the structural frame supporting the brick and the brick wall caused by the frame's lateral load deflection, the frame's differential temperature movement, the concrete frame's creep and shrinkage deformation, and moisture expansion movement of the brick.
4. Inadequately spaced, located, and constructed expansion-contraction joints.
5. Excessive deflection of brick walls resulting from wind pressure or suction, restrained temperature movement, or differential brick temperatures.
6. Lack of clearances between structural members and the brick masonry wall preventing differential movement.
7. Anchorage deformation performance of brick lateral support ties or anchors.
8. Wall flashings reducing lateral load or shear force support.

Nonperformance Factors

Nonperformance factors, relative to newer buildings and older buildings, can be placed into four groupings. These groupings are design, construction, materials, and ASTM Specifications. Maintenance typically is not a factor except for certain isolated conditions, and therefore is excluded from the grouping.

Design

Of all the nonperformance factors, the design of brick masonry walls is likely the most critical in determining whether or not walls perform. This stems from the design selecting the materials, determining compatibility among the various details and materials, determining the details of construction, and providing specifications for construction.

Nonperformance of brick walls has revealed the frequent design omissions which lead to later problems. The more usual nonperforming design aspects are

1. Inadequate evaluations of the wall as a whole relative to water penetrations, durability, structural behavior, and interfacing with other parts of the building.
2. Insufficient evaluation and determination of the required brick and mortar material properties.
3. Specifications not specifically defining material requirements or their testing, and not specifying the construction execution but rather relying upon "as-required" or "good workmanship" clauses.
4. Incomplete designs for flashings in terms of relying on typical details only and leaving the determination for the final required flashing construction details to workmen.
5. Neglecting to determine movement behaviors and vertical support requirements for exterior wall brick masonry, and not locating all required vertical and horizontal expansion-contraction joints.
6. Inattention to the wall's structural requirements concerning wind (pressure and suction) and support anchorages (ties, anchors, and lateral connections to the structural frame).
7. Not determining deflection behaviors of structural members or components, and their influence upon wall cracking or other distress to the brick wall.
8. Ignoring or not understanding recommended design and construction practices published by the Brick Institute of America in their "Technical Notes on Brick Construction."

Construction

Like design, the construction of brick masonry walls is also an important element in determining whether or not the walls perform. This results from the construction either complying with the design or not being built according to specified requirements. The best construction cannot result in a brick wall performing if the design is incorrect. And, if the construction is inadequate, a wall correctly designed similarly will not perform.

Construction related factors affecting brick wall behavior generally appear to involve the same items relative to nonperformance. Some of the more relevant construction factors are

1. Insufficient comprehension of the plans and specifications in terms of material requirements, and the details of the construction.
2. Lack of knowledge concerning how flashings and vapor barriers perform, and insufficient technical comprehension regarding the flashing details and their construction requirements.
3. Not installing or incorrectly installing the specified flashings.
4. Not installing or incorrectly constructing the specified expansion-contraction joints.
5. Head joints not being full and bed joints having furrows or voids, Figs. 7 and 8.
6. Inadequate or incorrect installation of wall ties and lateral support connections.
7. Improper tooling of mortar joints, and constructing mortar joints wider than 9.5 mm (3/8 in.).
8. Permitting 10.2 to 15.2 cm (4 to 6 in.) of mortar droppings to accumulate upon cavity wall flashings.
9. Allowing mortar bridgings to occur which bridge the collar joint or cavity at wall tie and floor locations permitting water to bypass flashings.
10. Not installing the weep holes specified for cavity walls.
11. Neglecting to strike mortar joints flush at locations where rigid insulation is to be adhered, and not installing the rigid insulation with tightly abutting joints.
12. Providing admixtures to mortar such as calcium chloride during cold weather construction.

FIG. 7—*Poor workmanship at building parapet wall resulting in open head and bed joints.*

FIG. 8—*Removed building wall wallette exhibiting poor workmanship in terms of nonfull head and bed joints, and mortar bridgings of cavity.*

13. Neglecting to employ cold weather construction procedures resulting in poor brick to mortar bond, or frozen mortar.

14. Failure to notify the designer when specified construction requirements or details cannot be implemented, or where their actual implementation will vary from that required.

Materials

The various materials used to construct brick masonry walls can, on occasion, be the cause of the wall's nonperformance. Material factors associated with the nonperformance of brick walls are

1. Mortars, complying with the type specified, which are not compatible with the wall's brick in terms of good bond thereby resulting in water penetrations.
2. Mortars using cements having high alkali contents which lead to efflorescence.
3. Bricks that do not comply with ASTM requirements.
4. Bricks which have initial rates of absorption above 30 g per min per 645 mm^2 (30 in.^{2}) and are not prewetted resulting in poor brick to mortar bond.
5. Concrete masonry backup having excess soluble salts which can lead to efflorescence in bricks complying with ASTM C 62, C 67, or C 216 requirements.
6. The use of calcium chloride in mortars which can promote efflorescence and poor brick to mortar bond, and cause corrosion of steel, wall ties, joint reinforcement and other embedded metal items.
7. The use of certain polymers in mortar for increasing mortar bond and strength which in the presence of alkalis release excessive chlorides causing ab-

normal and premature corrosion of metals, protected by special coatings or not.

ASTM Specifications

ASTM Specifications, in general, represent the overall consensus of industry, governmental, and user requirements for brick wall materials. However, in part, these ASTM requirements can be misleading in terms of not correctly informing the owner, designer, or contractor of their duties above and beyond fulfilling the ASTM requirements. Simply put, ASTM Specifications, or requirements, represent conditions of purchase—nothing more or less.

The ASTM factors relative to nonperformance of brick masonry walls are those not stated or those responsibilities undefined by ASTM which by tradition are assigned to the wall's designer and/or constructor. Some of the shortcomings of ASTM Specification C 62-81, ASTM Specification for Ceramic Glazed Structural Clay Facing Tile, Facing Brick, and Solid Masonry Units (C 126-82), ASTM Specification for Non-Load-Bearing Concrete Masonry Units (C 129-75 [1980]), ASTM Specification C 216-81, ASTM Specification for Mortar for Unit Masonry (C 270-82) and ASTM Specification for Grout for Reinforced and Nonreinforced Masonry (C 476-80) regarding nonperformance are:

1. The Specifications do not typically qualify that it is the responsibility of the designer or end user to determine the suitability of the material's integrated use into a structure for the specified conditions.
2. The Specifications neglect to clarify that neither the manufacturer nor the constructor is responsible for the in-service performance of the specified ASTM materials once they become part of an integrated structure providing the material complied with ASTM prior to being placed in the construction.
3. ASTM Specifications C 62-81 and C 216-81 imply that the saturation coefficient provides a measure of the brick's durability, when actual in-service use frequently indicates the *C/B* ratio is not a measure if the bricks are critically water saturated.
4. The Specifications fail to define or adequately provide data relating to the compatibility of mortars and bricks as affects bond, and hence resistance to water penetration.
5. The Specifications neglect to define that the Standards only relate to certain minimum qualities which must be satisfied at the time of purchase.
6. The Specifications fail to state that compliance with the Standards will not necessarily result in satisfactory performance of brick walls.

Nonperformance Evaluation Methods

Many evaluation methods are available to determine the cause of nonperformance. These methods range from visual investigations to chemical and petrographic analyses.

Visual Examinations

Visual examinations offer a means to either determine the cause of the nonperformance or to provide the initial information which leads to cause identification. Relative to building-related nonperformance, visual examination can consist of observations accompanied by notes and photographs, or a more comprehensive survey.

The detailed comprehensive survey provides the most desirable approach since when it is reduced to recording drawings it allows the investigator to readily view the building in its entirety. Study of recording drawings often reveals distress patterns associated with the nonperformance leading to defining locations or conditions where further evaluations should be made. In particular, the following comprehensive survey method records:

1. Abnormal and normal conditions of sealants (caulking) and deformations of sealants.
2. Abnormal and normal conditions of construction material adjacent to the brick, such as window frames, doors, mechanical openings, precast concrete, or metal siding.
3. Locations of efflorescence.
4. Separations or gaps between the brick and mortar about the brick perimeter.
5. Locations of stains or dampness.
6. Conditions of the mortar joints relative to size, tooling, shrinkage, cracks, and voids.
7. Cracks in head and bed joints or through the brick.
8. Behavior conditions at expansion-contraction joints.
9. Conditions of wall copings, all visible flashings, presence of weep holes, and wall vents.
10. Differential movements, or bulging or bowing behavior of the wall.
11. Structural related distress.
12. Unusual corrosion of exposed lintels or shelf angles, or other exposed metal items.

The detailed survey can be made from the ground using binoculars or telescopes. Or, the detailed survey may be made from swing stages lowered down the sides of buildings. Experience repeatedly indicates that close-up examination by the unaided eye provides the best visual data.

Physical Openings

Frequently, upon completing the visual examination, openings in the in-place construction are made to further examine certain conditions and obtain material samples for testing.

Openings are typically made either by using hammer and chisel or by employing power saws (dry or wet cutting depending upon the opening requirements). Openings provide information on

1. Fullness of head and bed joints.
2. Existence of voids, bed furrows, or separations throughout the entire mortar joint.
3. Presence of mortar bridgings and mortar droppings.
4. Presence of unusual amounts of water.
5. Manner in which flashings have been installed and their effectiveness.
6. Existence of wall joint reinforcement, ties, and other masonry anchors, and the manner by which they have been installed.
7. Conditions of embedded metals relative to corrosion.
8. Distress conditions of the brick.
9. General construction details regarding clearances between masonry and structural frame members or other conditions which restrain movements.

Openings to obtain wallettes are most often selected to be 40.6 by 40.6 cm (16 by 16 in.) or 60.9 by 40.6 cm (24 by 16 in.). Material samples for further testing can be obtained from the wallettes. In certain instances, when it is desired to determine the in-place moisture content of the brick, the brick should be placed in a sealed container immediately and tested that day or the next.

Water Permeability Tests

Field water permeability testing offers a tool not only to evaluate a brick wall's water permeability but also to provide a means for measuring both the materials and the workmanship used in the construction. The field water permeability tests presently used by several engineering firms, including the author's, basically follow the requirements of ASTM Test Method for Water Permeance of Masonry (E 514-74) with the exception that the test is conducted upon in-place walls.

A typical test setup is shown by Fig. 9. The 1.22 by 0.91 m (4 by 3 ft) test chamber mounted upon the brick wall is used to conduct the test at a 10 psf 0.48 kPa, (10 psf) pressure and water flow rate of 154.4 litres (40.8 gal) per hour through a spray bar at the chamber top. The pressure inside the chamber is measured by a water manometer and the flow rate by a flow meter. The apparatus depicted employs a closed water system. Water is pumped from a calibrated tank to the chamber and that water which does not enter the wall drains back into the tank. The chamber is mounted onto the wall using a gasket supplemented with sealants to ensure there is no water leakage between the wall's brick and the chamber.

Obviously, brick walls are not subjected to wind and rain of the magnitude developed by the test. However, if all tests are conducted at the same pressure and water flow, then a reference base can be developed to which actual brick

FIG. 9—*Field-adapted ASTM E 514 water permeability test.*

wall service performances can be compared. The test, as performed by the author's firm, typically is conducted according to the following criteria:

1. The test is started and conducted for one-half hour without any measurements being taken except the tank water level at the end of this period. This represents preconditioning and initial absorption of water by the wall.
2. One-half hour after the time the test is started, the water level in the tank is recorded. Throughout the test, the level of the water in the tank is recorded at each test half-hour.
3. The test is conducted for a minimum of four hours.
4. After four hours the test may be terminated providing the loss of water from the tank is approximately the same for at least the last three half-hour measurements.
5. The test is terminated after between four and seven hours once the last three half-hour tank readings indicate an equal loss quantity of water from the tank.
6. The test is terminated, regardless, at the end of seven hours and the rate of water loss is determined from that quantity lost from the tank in the last test hour.

The foregoing test procedure provides data to determine the rate of water loss per hour in either gallons per hour or liters per hour per 1.11 m^2 (12 ft^2) of wall. Comparison of test leakage rates to service performance indicates the following:

1. Walls having a test water loss rate of 0.71 litres (3⁄16 gal) or less per hour should be considered to be excellent in terms of service permeability.
2. A test water loss rate of 1.89 litres (1⁄2 gal) or less per hour should be con-

sidered as that which results from average good wall construction in terms of service permeability.

3. Walls having a test water loss rate of 1.89 to 3.78 litres (½ to 1 gal) per hour indicate questionable permeability, and walls which may leak during periods of high winds and rain.

4. Walls having a test water loss rate of 9.46 litres (2½ gal) or more per hour have high permeability and can be expected to leak during any rainstorm regardless of wind levels.

Fig. 10 represents a typical test water loss rate during the permeability test.

Field water permeability tests provide information on the brick wall's materials and quality of construction. The test results signify the following:

1. Low test rates of permeability, 1.89 litres (½ gal) per hour or less, indicate the wall's materials are of good quality, materials are compatible (good bond), and the workmanship is above average.

2. Test water loss rates of 1.89 to 3.78 litres (½ to 1 gal) per hour raise questions concerning the wall's materials and the construction workmanship.

3. Test water loss rates in the range of 5.68 litres (1½ gal) per hour indicate something is seriously wrong in terms of the materials used or the construction workmanship or both.

4. Test water leakage rates of 7.57 litres (2 gal) or greater per hour indicate the wall's materials or construction workmanship or both have serious problems.

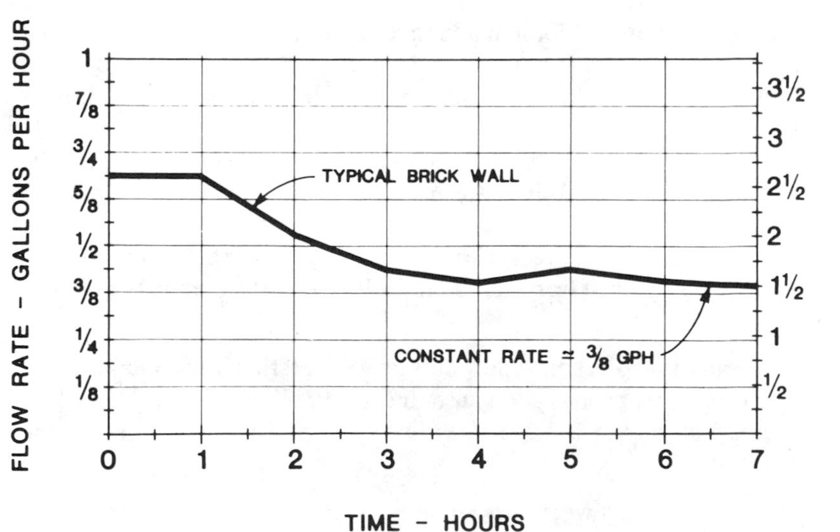

FIG. 10—*Modified ASTM E 514 field permeability test results.*

Wall Structural Tests

Structural tests of in-place brick walls or removed wall components are conducted to determine flexural or flexural bond strength, wall safety, deformations, effectiveness of lateral supports and wall ties, temperature gradients, and locked-in forces or stresses.

Tests for determining structural behavior can be conducted in a variety of ways. One common method is that shown by Fig. 11 where air evacuation (suction) is used to place a uniform loading upon the brick wall. The magnitude of the load is measured by using a water manometer. When conducting a uniform load test by air evacuation, or other methods, it may be necessary to make saw cuts in the wall to ensure the structural behavior is not influenced by the adjacent unloaded wall portions. Another consideration relative to the suction testing is to ensure the chamber box is capable of supporting the weight of the test area brick if a collapse develops during testing. Therefore, it is important to monitor wall deflection during testing and maintain a load-deflection plot. Assuming deflections are plotted on the graph abscissa, it is advisable to terminate the test if the load-deflection curve tends to become horizontal (large increases in deflection with slight change in load).

Brick surface temperatures can be readily measured using infrared equipment or thermocouples. Thermocouple measurements are most desired if a realistic understanding of the thermal gradient is required.

The determination of locked-in brick wall stresses, or forces generated by restraint movements (particularly temperature), likely represents the most

FIG. 11—*Structural load test using suction to determine wall's load-deflection behavior.*

sophisticated testing. Often it is only possible to mount mechanical or electrical gages on one surface. When single-surface gaging, rather than gaging on both faces of a brick wythe, is used, it requires pretest sample removal so the gaging can be calibrated against known laboratory axial and flexural forces.

Structural testing of brick walls requires previous engineering and testing experience to prepare the tests properly and interpret the results. Test results can be meaningless if the necessary experience is lacking, as the testing apparatus may influence the wall's strength.

Material Tests

Testing of materials provides a means to evaluate whether or not the materials themselves are related to the nonperformance. Some of the material tests commonly employed are

1. Testing unused and removed brick for compliance with ASTM Standards C 62-81, C 126-82, and C 216-81 for physical properties using ASTM C 67-81 sampling and testing methods.
2. Petrographic and chemical tests of mortars to determine their composition, and hence type.
3. Petrographic and chemical tests to determine the presence of certain substances (namely chlorides or other deleterious materials).
4. Petrographic examination of brick and mortars to evaluate if freeze-thaw damage has occurred.
5. Flexural tests on removed wall prisms to measure brick-mortar bond.
6. Physical tests to measure bricks' coefficient of thermal expansion, modulus of elasticity, and moisture expansion characteristics.
7. Chemical analyses to determine the chemical compositions of bricks, mortar, concrete masonry backup, and efflorescence salts to evaluate the source of efflorescence.

Testing of materials removed from an in-service brick wall requires appreciation of the differences that can develop between the used bricks and the same bricks never installed in a wall. Certain considerations are

1. Physical properties relating to strength and absorption are basically not affected by the brick being in service unless freeze-thaw damage has developed.
2. Testing for the brick physical properties of walls sustaining freeze-thaw damage requires selection of the in-service test brick from protected nonfailed locations.
3. Specific removal, handling, strapping, and packing procedures are necessary for removed prisms which are to be tested for flexural bond or other bond characteristics.
4. Determination of a brick's contribution to efflorescence requires tests on brick which have never been in service nor been in contact with the wall's mor-

tar or concrete masonry backup materials if the brick's chemical makeup is suspect.

Preventing Nonperformance

Technical knowledge and understanding currently exist to prevent nonperformance of brick masonry walls. This information, for practical purposes, is published by the Brick Institute of America in thier "Technical Notes on Brick Construction." That information source plus ASTM Standard Specifications should prevent nonperformance regarding materials, design, and construction.

Materials

More careful selection of the brick wall materials relative to each other and the environment where they will be used is required. Materials selection should consider

1. Brick's exposure to moisture, rain, and condensed vapor, with regard to water penetration, freeze-thaw durability, and efflorescence.
2. Bond requirements for the brick selected relative to the mortar type chosen, and if special laboratory mortar mix designs are required.
3. Influence of coatings or other brick surface treatments on the brick's ability to breath.
4. The materials in the mortar that can affect its bond to brick, contribute to efflorescence, or promote premature and abnormal corrosion of wall ties, anchors, or reinforcement.
5. The brick's physical characteristics pertaining to strength, absorption, initial rate of absorption, thermal coefficients, moisture expansion behavior, and potential for efflorescence.

Design

Proper design of brick masonry walls requires the designer to understand the materials, the type of wall to be used, factors influencing the wall's performance, and integration of both the wall's various components and the wall with the building. General considerations are

1. Evaluate and select all materials in the wall regarding compatibility, serviceability, durability, strength, deformations, movements, and the environment.
2. Detailed studies of the brick wall's movements, movement restraint offered by the supporting structural components, and differential structural movement of the supporting frame relative to locations of brick expansion-contraction joints.
3. In-depth designs concerning the wall's resistance to water penetrations and all the requirements necessary to maximize water penetration resistance.

4. Design of and selecting materials for all flashing requirements, not just the typical conditions, to ensure continuous effective flashing at structural members, intersecting walls, variations in wall thickness, penetrations (electrical, mechanical, and plumbing), and changes in wall direction.

5. Providing walls with the ability to breath through their exterior face, and from within cavities for glazed brick walls, parapets, and free standing walls.

6. Planning for tolerances and making space provisions to permit interfacing of all the wall components and the structural frame.

7. Selecting nominal flexural deflection limitations for the brick wall itself (wall span divided by 1000) or of support beams (beam span divided by 1200) to prevent wall cracking unless a more refined stress analysis is required to evaluate cracking.

8. Employing specifications which define all materials and installation requirements, which do not require contractors to select any materials or perform design functions, and which do not require contractors to interpret or identify codes relative to design and materials.

Construction

Construction of brick masonry walls for best performance requires comprehension of the specified construction requirements and construction practices which result in quality construction. General aspects are

1. Comprehension of all the plan and specification requirements.

2. Construct brick masonry walls in accordance with the plans and specifications without contractor-selected variations.

3. Understand all flashing requirements prior to starting construction, make all flashings continuous without interruption throughout, and provide flashing end dams at points of flashing termination.

4. Ensure all head and bed joints are full, and joints are properly tooled.

5. Employ means to prevent mortar droppings from depositing on flashing, and prevent the development of mortar bridgings.

6. Install all expansion-contraction joints and do not permit mortar within these joints.

7. Rigid cavity insulation should be installed with tightly abutting joints in one plane where specified.

8. Vapor barrier installation should be continuous without voids or interruptions.

9. Joint reinforcement, ties, and anchors should be solidly embedded in each wythe's mortar.

10. Prevent mortar droppings and bridgings from filling planned clearances between walls and structural members.

11. Follow and implement cold weather construction procedures and do not use cold weather additives in the mortar.

12. Construction brace brick walls, using an appropriate safety factor, for winds of at least 80 km/h (50 mph), or greater if required by local conditions.

13. Clean walls only with the specified materials.

14. Report immediately any recognizable problems regarding errors in the plans or specifications, interfacing or congestion difficulties, and conditions hindering or preventing the required construction from being accomplished.

Owner

The owner, typically not considered, plays an important role in assuring brick masonry walls perform as intended. The owner should

1. Select architects and engineers who are qualified and have the necessary brick wall design experience to prepare in-depth and complete construction plans and specifications.

2. Use contractors and material suppliers noted for the quality of their work and those firms which employ sound planning, scheduling, organization, and project coordination.

3. Require qualified responsible inspection during the construction by either the designer or other firms where the inspectors have extensive experience in all phases of brick masonry design and construction.

Summary

This paper has discussed brick masonry walls regarding nonperformance categories and factors associated with nonperformance. Likewise, the paper has reviewed methods of evaluating nonperformance and recommendations to prevent nonperformance.

The topics and items selected for discussion herein are those the author's experience indicates are important for designers, contractors, material suppliers, and owners to understand if required performance is to be achieved. Certainly, other aspects than those reviewed can play a role in nonperformance. However, the major item to focus upon is responsibility—the responsibilities each member of the entire building team must assume if nonperformance is to be prevented. Towards this end, this paper has attempted to discuss some of these overall responsibilities.

References

[*1*] "Ceramic Glazed Brick Facing For Exterior Walls," Technical Notes on Brick Construction, No. 13, Brick Institute of America, McLean, VA, May 1962.

[*2*] Anderegg, F. O., "Efflorescence," ASTM Bulletin No. 195, Oct. 1952.

[*3*] Young, J. E., "Backup Materials as a Source of Efflorescence," *Journal*, American Ceramic Society, Vol. 40, No. 7, July 1957, pp. 240–243.

[*4*] "Efflorescence, Causes," Technical Notes on Brick Construction, No. 23, Brick Institute of America, McLean, VA, Dec. 1969.

Clayford T. Grimm[1]

Durability of Brick Masonry: A Review of the Literature

REFERENCE: Grimm, C. T., **"Durability of Brick Masonry: A Review of the Literature,"** *Masonry: Research, Application, and Problems, ASTM STP 871*, J. C. Grogan and J. T. Conway, Eds., American Society for Testing and Materials, Philadelphia, 1985, pp. 202–234.

ABSTRACT: The paper summarizes the engineering literature published in English since 1900 on the durability of brick masonry. Destructive agents, mechanics of destruction, porosity, freeze-thaw resistance, mortar properties, florescence, environment, architectural engineering, brick specifications, construction, and maintenance are discussed. Bibliographies with a total of 228 entries are provided on the durability of brick, masonry, brick masonry, mortar, and florescence. Manufacturing quality control is not considered.

KEY WORDS: brick, construction, corrosion, durability, efflorescence, failure, freeze-thaw, masonry, maintenance, mortar, porosity, specifications, walls

Vitruvious [*168*],[2] a Roman historian at the time of Christ, wrote, "Whether the baked brick itself is very good or faulty for building, no one can judge off-hand . . ." To this day there is no infallible method for detecting in advance whether or not a brick will be durable [*28,44,45,140*]. Although attempts to solve the problem by laboratory testing began in 1828 [*31*], the problem persists because there is no consensus on the definition of exposure severity, no standard classification of types of deterioration, no agreement on how much in-service deterioration is tolerable, and no certain correlation between a measured physical property and degree of deterioration.

The present ASTM standard specifications for brick are directed at frost resistance and virtually ignore other destructive agents well known to affect durability. Even the frost resistance requirements are grossly inadequate,

[1]Consulting architectural engineer and senior lecturer in architectural engineering, University of Texas at Austin, Austin, TX 78758.

[2]For the reader's convenience the italic numbers in brackets refer to citations found in the Bibliography appended to this paper.

providing for the acceptance of about 23% of brick known to be nondurable. The current standard freeze-thaw test is expensive, time consuming, unrealistic, and incapable of certain identification of nondurable brick.

Behavior can be predicted only when environment is exactly defined [*66*]. McBurney [*69*] suggested in 1931 that development of a standard test method for brick durability should await the standardization of climate and exposure. Some have thought that the only absolute test of weather resistance is 100 years or more of exposure [*149*]. However, Robinson et al [*91*] suggested that a reliable test for brick durability would be exposure for five years. One might assume that one or the other suggestion is wrong, but it is more probable that both are wrong. The problem is further complicated by the fact that brick masonry is an assemblage of two dissimilar materials, brick and mortar, which individually may be durable but which may not combine to provide a durable composite [*139*].

Durability is a function of materials, design, construction, environment, and maintenance. Durability may be defined as the ability of a material to remain serviceable with prudent maintenance during a normal life span in the intended environment; that is, to retain physical, chemical, and visual performance characteristics within reasonable tolerances for an economic life expectancy when properly designed, built, and maintained [*152*]. The National Bureau of Standards made a detailed study of 250 cases of efflorescence and wall disintegration, and concluded that all but one or two could have been prevented by either proper design, construction, or maintenance [*210*].

The following physical properties are important to durability of masonry: porosity, pore size distribution, water absorption, saturation coefficient, capillarity, drying rate, rate of water absorption, water permeability, air permeability, salt content, and tensile, compressive, and shear strength [*96,172,217*].

Frost resistance is a property not only of the design, construction, and materials exposed to frost but also of the condition of exposure [*44,45*]. If the properties of material and environment are well defined, it is still necessary to define what changes, if any, are acceptable in a brick to be considered frost resistant; that is, "mellowing" is acceptable and "weathering" may not be. Slight color change due to some loss of sand finish or fine texture or due to minor dirt accumulation are usually acceptable. Major changes are not. Brick masonry may crumble or spall from salt crystallization, freeze-thaw, or structural stress concentration, and there is currently no means to distinguish among them except the considered judgment of an experience observer [*87*]. However, introspective subjectivism, not to say whim and caprice, is often the present arbitary durability performance criterion use by some building design professionals.

The search for truth on this subject has continued for many years. In 1931 Parsons [*80*] annotated a bibliography on the weathering of structural clay products with 63 references. McBurney [*69*] provided a review of the literature prior to 1931. Butterworth [*33*] provided a review of the 1900 to 1951 literature with 278 entries. In 1964 [*35*] he updated the work with a critical review of the work published in 1952 to 63 with 78 citations.

This paper provides a synthesis of the literature published in English since the turn of the century on the durability of brick masonry, including destructive agents, mechanics of destruction, relevant material properties and test methods, florescence, environment, brick specifications, architectural engineering design considerations, construction, and maintenance. Materials manufacturing quality control is not considered. The literature on brick masonry durability is scattered through publications devoted to engineering, architecture, construction, and materials science. The subject of brick masonry durability is outlined in Appendix I, and the list of references includes 228 sources.

Destructive Agents

Johnes [*64*], McIntyre [*77*], Brightly [*114*], Kessler [*133*], Dorey [*119*], McBurney [*134*], and Frohnsdorff and Masters [*121*] have all contributed to a list of factors which affect the durability of brick masonry. Churchill [*152*] thought frost action, that is, freeze-thaw of absorbed water, is the most important of the several destructive agents. However, Lawrie and Milne [*215*] and Anderegg [*194*] thought that florescence caused most masonry disintegration, but Butterworth [*29,44,45*] thought such destruction was rare.

Water may enter masonry from the ground, by condensation, by rain entry through cracks, or by facial absorption of water [*139*]. Leaks may occur in roofs, gutters, down spouts, parapets, flashing, and sealant joints. Cracks are the principal means by which water enters masonry, and the effect of cracks on durability is well recognized [*114,132,136,138*]. The subject of masonry crack control water permeance is outlined in Appendix II and in Appendix III.

Water which enters masonry may contain salts, or the water may dissolve salts within the masonry. When the water evaporates, the salt crystalline residue is called "florescence." Salt deposited on the face of the masonry is a stain called "efflorescence." If evaporation occurs inside the face of the masonry, the salt crystalline residue is called "cryptoflorescence" [*207*]. Efflorescence may be an aesthetic problem, but does not affect durability [*205*]. It may, however, be indicative of cryptoflorescence, which can be destructive [*50,133,173,203,211,215,223,224*]. Brard in 1828 and others since [*69,91*] proposed salt crystallization as a test method for brick durability.

Mortar is subject to the same agents of deterioration as is concrete. Pollutants in the environment and contaminants in the masonry accelerate mortar deterioration. Atmospheric pollutants include carbon dioxide, sulfur dioxide, hydrogen sulfide, salts, ammonia, nitrate, nitrite, and particulate matters [*142*]. Excessive salts are often found in earth adjacent to foundations, retaining walls, and pavements. Contaminants include salts, oxides, and carbonates [*157*]. Brick may be significantly affected by salts and ice but not significantly by the other pollutants and contaminants. However, improperly protected metals embedded in masonry such as anchors, ties, reinforcement, lintels, and shelf angles may be greatly affected [*157*]. Their corrosion can cause masonry to crack and, therefore, affect masonry durability.

Winkler [*147*] describes the effects of bacteria, algae, fungi, lichens, mosses, plant roots, animal boring, and bird droppings on masonry. Lichens, mosses, fungi, and algae may be aesthetically objectionable, but there is no evidence that they have any significant affect on durability [*69,109*]. Plant roots do affect durability of masonry.

Mechanics of Destruction

The imposition of external loads can induce excessive stress and cause cracking. Restraint of internally generated dimensional change can exert pressure capable of causing cracks. Cracks can lead to ultimate destruction. Brick masonry changes dimensions for several reasons. Jessop and Baker [*131*] provide a 1980 annotated bibliography on moisture, thermal, elastic, and creep properties of masonry. Thermal expansion is an example of expansion of brick. Freezing of water in pores is an example of expansion within brick. Expansion of masonry is an inherent property, which should be accommodated in design. Excessive unaccommodated expansion of masonry is evidence of inadequate design. Excessive unaccommodated expansion within brick or within mortar is evidence that the materials were subjected to an excessively severe exposure or that the materials *per se* have inadequate durability. Measurable expansion of brick is no evidence of nondurability of brick [*159*].

Brick have a net long-term moisture expansion which increases with time lapse. If restrained moisture expansion is 0.02% and if the elastic modulus of brick is 20 684 MPa (3×10^6 psi), the stress is about 4.14 MPa (600 psi). Restrained thermal expansion may also be responsible for high stress. Consider a temperature change of 27°C (80°F) in a brick having a coefficient of thermal expansion of 1.67×10^{-6} m/m · °C (3×10^{-6} in./in. · °F). If the elastic modulus of the restrained brick is 20 684 MPa (3×10^6 psi), the thermal stress is about 720 psi (4.96 MPa). The coefficient of thermal expansion of brick is increased 20% to 30% or more by freeze-thaw cycles [*103*].

Suppose that water in a saturated brick fills a small pore space and that the temperature of the brick is rapidly increased from 70°F (21°C) to 150°F (66°C), due to solar radiation and diurnal temperature rise. If the coefficient of thermal expansion of the clay is 1.67×10^{-6} m/m · °C (3×10^{-6} in./in. · °F), the thermal strain in the clay is $(150 - 70)\, 3 \times 10^{-6}$ or 0.024%. However, the coefficient of thermal expansion of water at 66°C (150°F) and at atmospheric pressure is about 118×10^{-6} m/m · °C (212×10^{-6} in./in. · °F) [*119*]. The water expansion is, therefore, about 1.69%. If the water is not free to escape due to cryptoflorescence, a surface coating, or other reasons, the brick is under very high internal stress.

Winkler [*147*] describes theoretical osmotic pressures which may develop in masonry materials. These pressures may be as high as 10.3 MPa (1500 psi). McIntyre [*135*] had discounted the effect of osmotic pressure.

Considering 14 different salts at 2% concentrations and at temperatures of 0 to 50°C (32 to 122°F), the crystallization pressure is said to range from 5.06 to 66.3 MPa (735 to 9611 psi) [*147*]. Pressure increases logarithmically with salt concentrations.

Hydration of sodium sulphate causes an expansion of about 82% [*135*]. Bonnell and Nottage [*195*] measured the pressure due to hydration of salt in pores. Winkler [*147*] points out that the full evaluation of salts in very narrow capillaries is very difficult. However, he mentioned that the pressure exerted by the hydration of salt increases with reduced temperature and increased humidity. At low temperature and high humidity, hydration of plaster of paris to gypsum could exert a pressure of more than 206.8 MPa (30 000 psi).

The coefficient of thermal expansion of salt may be more than twice that of masonry. A rise in temperature could cause considerable pressure against the walls of a pore filled with salt.

If there are varying opinions as to the precise mechanisms of masonry disintegration due to weathering, there is little dispute as to the root cause [*148*]. Ritchie [*86*] lists the following factors affecting frost damage: (1) moisture content; (2) rate of temperature change during freezing; and (3) direction of freezing.

Butterworth [*44,45*] provides an excellent summary of the various theories of freezing in a porous body. Thomas [*146*] provides a convincing description of the mechanics of freezing in a porous material. Gardner [*123*] explains freezing in masonry as follows:

> Freezing advances into the masonry in a plane parallel to the exposed surface. The freezing of water reduces potential and induces migration of water to the warm side of freezing plane. Upon freezing, the release of the heat of crystallization acts to retard the advance of the freezing plane. So long as the advance of the freezing plane is retarded in that way, water continues to migrate to the freezing plane where it freezes, causing ice-lensing. Growth of the ice lens produces expansion far in excess of the normal expansion of water upon freezing.

Reduction of stress due to freezing of a saturated, or nearly saturated, porous body depends on water being able to move away from the freezing zone into an unfrozen zone or by extrusion from pores from the warm side [*32*]. Frost resistance is, therefore, related to the rate at which water can flow through the body. Rapid freezing prevents extrusion and increases disintegration [*21,33*].

The formation of ice crystals restrained in the pores of a saturated brick can exert pressures of more than 213.7 MPa (31 000 psi) at −22°C (−8°F). However, large pore diameters permit outward drainage and escape of water as the ice front advances through the brick, thus reducing restraint to the ice and, therefore, the pressure exerted by it. However, Stanford [*94,96*] found that brick, which are frost resistant by the conventional laboratory freeze-thaw method, have in fact suffered internal damage after only a few cycles.

Freezing causes expansion within masonry pores and also expansion of masonry. The most likely freezing expansion of stiff mud brick is about 0.039% with a standard deviation of about 0.0025% [*55*].

Porosity

Early attempts to define porosity and to describe pores were made by Washburn [*101*] and subsequently by others [*18,99,100*]. In general, today porosity is understood to mean the ratio of the volume of air contained within the boundaries of a dry material to the total volume (solid matter plus air), expressed as a percentage. Total effective porosity is determined by mercury intrusion according to ASTM Standard for Chemical, Mass Spectrometric, and Spectrochemical Analysis of, and Physical Tests on Beryllium Oxide Powder (C 699). The internal cavities, called pores, may be open or closed. Few brick contain more than 2% by volume of closed pores [*92*]. Brick lack homogeneity and have many different sorts of internal structure (*100*].

Water absorption by weight is an indication of porosity and is easily determined as the weight of water absorbed after submersion for 24 hours in cold water as a percentage of the dry weight of brick according to ASTM Standard for Sampling and Testing Brick and Structural Clay Tile (C 67). Water absorption is also determined by submersion for five hours in boiling water. The ratio of the 24-hour cold water absorption to the five hour boiling water absorption is called the saturation coefficient or the *c/b* ratio. It is an approximation of the ratio of easily filled pore space to total pore space. McBurney [*69*] credits Hirschwald (1912) and Kreuger (1923) with the development of the theory that as water expands ten percent in freezing, there must be at least 10% unfilled pore space present in brick to provide the needed space, otherwise disintegration will occur.

The rate of water absorption by bricks is an indicator of pore size. Large pore size is associated with greater durability. The initial rate of absorption, called *IRA* or suction, is measured as the weight of water absorbed per unit time, per unit surface area upon immersion in various depths of water. ASTM C 67 requires measurement of the weight of water in grams absorbed by brick in one minute per 193 cm^2 (30 $in.^2$) while immersed in 3 mm (1/8 in.) of water. Suction is reduced at lower temperatures [*54*], an increased as the square root of the immersion time lapse [*61,85*]. IRA is reduced by moisture content of brick at time of test [*30*] and by water repellant treatment of the brick surface [*85*].

The historical description of brick as "vitrified," "hard," "medium," and "soft" were quantified by McBurney [*68*] as given in Table 1.

Douty and Beebe [*56*] determined water absorption of a variety of brick by

TABLE 1—*McBurney's historical description of brick* [68].

Term	5 h Boiling Water Absorption, %	Ratio of 5 h Cold Water Absorption to 5 h Boiling Water Absorption
Soft	above 20	0.85
Medium	12 to 20	0.75
Hard	5 to 12	0.65
Vitrified	0 to 5	0.55

five methods as a function of time from ten minutes to 110 days. McBurney [*72*] discussed the effect of testing technique on water absorption test data. McBurney [*68*] tested the water absorption of many brick samples by a variety of methods and compared the results to in-use performance. In discussing that paper Mr. D. W. Kessler stated, "The weathering of masonry involves so many different factors that no single test can ever be of much value."

By 1935 McBurney [*71*] concluded that a combination of strength, water absorption, and *c/b* ratio provides the most accurate prediction of freeze-thaw resistance found to that time. However, saturation coefficient alone is not reliable in predicting freeze-thaw resistance of brick [*35,68,93*]. Products having saturation coefficients as low as 0.65 may lack frost resistance, and others having an index as high as 0.97 have proven frost resistant. In-mixture of relatively coarse-grained combustible material in the clay prior to burning results in lower saturation coefficients of bricks but no change in frost resistance [*48*].

Sandford [*93*] and Litvan [*66*] found a variation in saturation coefficient within the same brick and among brick from the same lot. He found that the *c/b* ratio of seven brick from the same production lot ranged for half brick from 0.7 to 0.88. The *c/b* ratio decreases with decreased specimen size.

There is a significant difference between water absorption values determined by different laboratories on the same batch of brick. In the case of boiling water absorption some laboratories produce consistantly higher results from others [*82*]. Stull and Johnson [*98*] found the difference between porosities of different sections of individual bricks ranges from 11.6 to 48.42%.

Butterworth [*30*] provided data on the absorption, porosity, and density by several methods on several kinds of brick. Stedham [*97*] developed a rapid test method for determination of brick water absorption. McBurney [*70*] provided water absorption test data on brick from 225 plants. McBurney [*74*] plotted the group average water absorptions, saturation coefficients, and strength against one another and found smooth and regular curves.

Grimm and Houston [*60*] compared water absorption of whole brick removed from a wall with that of the original brick and found no significant effect of mortar contact on total water absorption or *c/b* ratio. However, mortar contact significantly reduced IRA. Grimm and Houston [*60*], also found that plant siliconing can result in improper grading of bricks, if the silicone is not removed prior to water absorption testing.

Beginning in 1952 with Carlson [*48*], several investigators have claimed to show a relationship between pore-size distribution and frost resistance [*26*]. There is some evidence that the number of laboratory freeze-thaw cycles at which brick specimens fail is directly proportional to average pore diameter. Brownell [*26*] indicates that typical frost resistant brick have an average pore diameter of 1.6 μm, while brick having low frost resistance have an average pore diameter of 0.24 μm (1 μm = 0.01 mm = 0.0004 in.). Based on very limited data, Winkler [*147*] suggested that a clay body will be durable if at least 30% of porosity consists of pores with an equivalent diameter greater than 0.8 microns. Astbury and Vyse [*20*] developed a method for study of pore size distribution which provides a measure of pore geometry at pore diameters down to 0.05 μm. The method provides more information than the mercury porisimeter.

McBurney [*73*] found that air permeability, number of capillaries per unit area, and average area of capillaries were so poorly correlated with durability that no limits could be established.

Freeze-thaw Resistance

The language with regard to the weathering of brick masonry is colorful; for example, "spalling, pitting, popping, cracking, chipping, exfoliating, and staining" to which for glazed brick can be added "peeling, crazing, crawling, shivering, and craw footing." Within these there are many subclassifications.

Attempts have been made by visual observation to quantify disintegration due to weathering, but the subjective nature of a visual criterion has led to its abandonment [*23,27,67,81,103*]. Balwin [*21*] thought that as few as one spalled brick in 50 may be sufficient to spoil the appearance of masonry. Some have regarded a ten percent weathering failure rate as catastropic.

Accelerated laboratory tests for durability of brick date back to 1828. A form of laboratory freeze-thaw test that was standardized in 1887 is still in use today with minor variations [*31*]. The current standard freeze-thaw test for brick (ASTM C 67) involves soaking units in water at 24°C (75°F) for four hours and freezing them at −9°C (16°F) for 20 h, while immersed in 13 mm (½ in.) of water. After each fifth cycle the units are air dried for 40 hours at room temperature. Weight loss after 50 cycles is the failure criterion.

McBurney [*70*] tested 480 brick from 255 plants and proposed criteria for identifying those brick which would endure a freeze-thaw test. Those criteria with slight modification are used today in ASTM specifications; that is, water absorption, saturation coefficient, and strength. Conformance to the ASTM standards for brick is no guarantee of durability [*40*]. The present ASTM durability requirement would accept as many as 22.8% of nondurable brick and reject as many as 31.5% of the durable brick [*91*].

McBurney and Johnson [*75*] half buried 271 deaired brick from 13 manufacturers in soil in Washington, D.C., for five years, during which there were 258 natural freeze-thaw cycles. Brick were said to have "passed" the test if spalls on faces, edges, or corners did not exceed one sq in. Of the 271 brick, 219 were classified under ASTM specifications as grade SW; that is, suitable for exposure to severe weathering. Of the 219 SW brick, 25 did not "pass" the exposure test; that is, there was an 11% failure rate for the best ASTM grade. Of 35 grade MW brick, 10 did not "pass"; that is, there was a 29% failure rate. It was concluded that a required compressive strength of not less than 82.7 MPa (12 000 psi) or a five hour boiling water absorption of not more than 12% would practically eliminate the possibility of failure to withstand 75 cycles of laboratory freezing and thawing. However, these specification requirements would eliminate more than half the satisfactory building brick produced in the United States.

Davison [*53*] subjected several types of brick to 1500 cycles of laboratory freeze-thaw tests and in outdoor exposure to about 1000 freeze-thaw cycles while bedded one inch in a sand-gravel mixture. Some extruded brick which survived the laboratory test with no visible change failed dramatically on the exposure site. Some brick which conformed to the requirements of the purchase specification (ASTM) for absorption and strength failed the laboratory freeze-thaw test and performed well in outdoor exposure, while other brick with high saturation coefficient remained in excellent condition in laboratory freeze-thaw tests and failed in outdoor exposure.

In brick substantially free of defects, the number of cycles of freeze-thaw required to cause failure increases as air permeability and mean effective capillary radius increase, and when the saturation coefficient, porosity, and number of capillaries decrease [*98*]. There is no evidence that cored brick are less durable than solid brick of the same clay and degree of firing [*19*]. Richmond and McBurney [*84*] found freezing to −29°C (−20°F) much more destructive than freezing to −7°C (20°F). In some materials all the water is frozen at −20°F (−5°C), while with others there is an appreciable amount of water still unfrozen at −29°F (−20°C) [*24*].

Rate of freezing rather than temperature level is said to be the effective factor influencing disintegration. Llewellyn and Butterworth [*67*] found that a rapid rate of freezing was more destructive because it did not allow water to extrude from the pores. Thomas [*146*] found that one material may have superior frost resistance to another at one rate of cooling; however, the reverse may be true at another cooling rate.

Foster [*57*] recognized the need for laboratory freeze-thaw tests which more nearly simulated in-service environmental conditions by degree of saturation and direction of freezing. McBurney [*73*] believed that a more severe test than

50 cycles of laboratory freeze-thaw should be required for a waiver of the physical requirements of the ASTM specifications. However, Christensen [*115*] thought the conventional freeze-thaw test to be "too severe"; that is, more severe than exposure in real walls. The omnidirectional freezing of brick in the ASTM test method is more severe than the unidirectional freezing which occurs in a brick in a wall.

In the ASTM C 67 freeze-thaw test, brick are at a much higher moisture content when frozen and the rate of freezing is greater than would typically prevail in a brick in a wall [*140*]. Because of this difference some brick with good service records perform poorly in the ASTM test [*88*]. Sanford and Fredholm [*95*] proposed a new unidirectional freezing test method; however, they found that the method was not equally applicable to all types of brick.

Palmer [*79*] speculated that 100 laboratory freeze-thaw cycles is equivalent to 30 years of exposure to severe climate in an unprotected wall. Parsons [*81*] found three laboratory freeze-thaw cycles to be the equivalent of one year of natural weathering for a brick with stretcher face exposure in a wall in New England.

Palmer and Hall [*79*] noted that 100 freeze-thaw cycles increased absorption and reduced compressive strength. Tests on the strength of 14 different types of bricks after various periods of burial in earth up to 34 years indicate generally a slight but significant decrease in strength, probably due to frost action [*47*]. However, Davison [*53*] found an increase in compressive strength after 1500 cycles. The National Bureau of Standards [*7*] found that repeated freeze-thaw cycling and drying did not substantially change bulk volume and absorption. Weight, solid volume, permeability, and capillary radius were found to increase, and the porosity, density, and number of capillaries decreased [*7*]. The Bureau [*8*] reported an increase of about 1% in the ratio of 48-h cold water absorption to total pore volume for each laboratory freeze-thaw cycle. Butterworth [*42*] found that exposure to weather for 29 years reduced the rate of absorption but that total water absorption was only slightly changed.

Frost dilatometry (measured as the irreversible expansion of brick which occurs after five cycles of freeze-thaw) is said to be a better indicator of frost resistance than saturation coefficient [*41*]. But subsequent work has failed to provide adequate correlation between frost dilatometry test results and in-service durability [*43*]. The magnitude of residual expansion due to frost is strongly dependent on the degree of water saturation [*96*]. Watson [*104,105*] believed that distension (permanent set subsequent to freezing) holds promise of providing an improved criterion for predicting durability.

In 1973 Litvan [*66*] related frost durability to pore surface area of the burned clay body. He explained that no fully reliable test for evaluation of frost resistance has been developed, because the mechanism of frost action is not fully understood.

Mortar Properties

The coefficient of thermal expansion for mortar and brick may be about the same or may be three times higher for mortar than for brick or twice as high for brick as for mortar [*156*]. The differential thermal strain between surface clay brick and masonry cement mortar could be as much as about 0.04% for a temperature rise of 55°C (100°F). It can be shown that tensile stress thus produced in the mortar could exceed its strength, producing vertical cracks in bed joints [*135*]. Palmer and Hall [*163*] found that 50 cycles of laboratory freeze-thaw testing may not disintegrate bricks or mortar, but may break the bond between them. They point out that bond failure in masonry is more apt to occur in head (vertical transverse) joints than in bed (horizontal transverse) joints.

Type of mortar has a marked influence on the durability of masonry units [*170*]. Dense mortar may accelerate weathering of brick, because the drying rate from brick may exceed that of mortar, in which case salts can be drawn from the mortar into the brick and crystallize to cause damage [*215*]. Palmer [*164*] found the durability of mortars to be better for those having 24-hour cold water absorptions in the 6 to 12% range.

Autogenous healing is the capacity of high-lime mortars to reconstitute themselves by filling cracks. The process involves the carbonation of calcium hydroxide to form calcium carbonate. Calcium hydroxide formed during hydration of cement is more soluble in cold water than in hot water. The process is aided by cycles of cold, rainy weather and warm, dry periods [*171*]. If the calcium carbonate forms in cracks, it is beneficial autogenous healing. If it forms in brick pores it may be detrimental cryptoflorescence.

Air entrainment increases the durability of mortar [*174*]. Cement-lime mortars having 10 to 15% air entrainment have significantly greater durability than those containing 4 to 7% air entrainment. Air content above 15% does not improve mortar durability [*174*].

Mortar disintegration due to freezing while plastic (prior to initial set) is very similar to that caused by efflorescence, and it is very difficult to distinguish between the two [*173*]. However, the deleterious effects of freezing mortar prior to set has sometimes been exaggerated [*185*]. Mortar frozen for one day while plastic has an average strength at an age of one year equal to 99.2% of that of the same mortar unfrozen. Freezing mortar while plastic for seven days reduces the one-year compressive strength only about 10% [*184*].

Straight lime-sand mortars were used in brick masonry prior to 1980, when portland cement became commercially available in quantity. Those mortars were of very low strength but were generally durable when used in very thin mortar joints which developed good bond and low water permeance [*171*]. However, increased lime content reduces mortar durability [*122,132, 174,187*], but the durability of masonry in walls protected by coping at the top and a dampproof course at the base was not affected by mortar type used during a 16-year exposure in Washington, D.C. [*178*].

Florescence

The degree of efflorescence has been quantified in British Standard BS 3921 [*17*]. Serious efflorescence is defined as a heavy deposit covering 50% or more of the exposed face with surface powdering, flaking, or both, tending to increase with repeated wettings. Heavy efflorescence is the same as serious efflorescence without powdering or flaking. Moderate efflorescence covers 10% to 50% of the exposed face with no flaking or powdering. A thin deposit covering less than 10% of the exposed face is said to be slight efflorescence. No perceptible deposit is termed nil efflorescence.

The two conditions necessary to formation of florescence are the presence of soluable salts in the masonry and water permeance into the masonry [*210,218*]. Salts may be present in the brick, stone, concrete masonry, or mortar materials or may be carried into the facing by ground water or polluted rainwater. Hydrochloric acid used in cleaning masonry may react with potassium or vanadium to cause florescence [*203*]. Water which has percolated through earth may contain florescing salts. Florescence may result from the storage of saltfree brick directly on earth.

The chemical composition of typical efflorescence is listed by Conner and Okerson [*173*] and by Ritchie [*223*]. The most common florescing salts are white or grey sulfates of sodium, calcium, magnesium, aluminum, and potassium [*203,204*]. Magnesium sulfate (epsom salts) is nearly always the agent responsible when failure results from serious florescence [*205*]. Carbonates or silicates of calcium, sodium, and potassium occur less frequently.

Vanadyl salts produce green or yellow stains. Vanadic acid causes brown stains. Rarely, iron oxide stains develop on light colored, fire clay brick with black coring. More rarely chlorides, nitrates, and salts of chromium may cause florescence. Presence of chlorides and nitrates are usually indicative of external contamination [*210*]. Any water soluble salt which finds its way into masonry could cause florescence. Of all these salts, probably calcium sulfate (gypsum) is the most troublesome, because it occurs frequently and is difficult to remove due to limited solubility [*203*].

It is usually not possible to correlate the kind of florescing salt with its source; however, X-ray diffraction and chemical analysis of anions and cations can indicate a most probable source [*203*].

Calcium and vanadyl sulfates and manganese oxides indicate brick as the probable source. Butterworth [*206*] reported wide variations in the soluble salt content in brick of any one make, from unit to unit in any one kiln, and from time to time. Conditions conducive to cryptoflorescence are large concentrations of soluble salt in a brick having an absorption of about 6% to

12% [*203*]. Ritchie [*223*] found brick having a high initial rate of absorption more prone to efflorescence.

Mortars often contain the seeds of their own destruction. Mortars containing more than 0.6% sodium or potassium oxides (soda equivalent) were consistently found by Conner and Okerson [*173*] to disintegrate in brick masonry walls in the New Jersey area.

Gypsum found as an impurity in brick or added to cement at the mill to control set may contribute to heavy efflorescence of masonry [*216*]. The concentration of alkali may release sulfate ion from ettringite in set cement paste, allowing it to migrate to produce florescence. Lime set free during hydration of portland cement, or lime in the mortar, may migrate to the surface and carbonate to form calcium carbonate efflorescence [*216*].

Portland cement is a probable source of sodium and potassium carbonate. "Free-alkali" solution (i.e., sodium and potassium hydroxides, NaOH and KOH) in cement can migrate from mortar to bricks, which are often more porous than mortar. These hydroxides may appear on the surface of bricks after evaporation of water from the bricks. The hydroxides then carbonate to form sodium and potassium carbonates [*173,203*]. There is evidence that the amount of portland cement in mortar influences efflorescence [*223*]. Portland cement may be present in mortar, concrete masonry backup, or in structural concrete supports.

Lime and washed sand in mortar do not contribute to significant efflorescence [*180,194,203*]. Good sand gradiation minimizes mortar porosity and reduces water permeance and, therefore, florescence [*216*].

Interaction of brick and mortar is illustrated by Laurie [*215*] who points out that sulfuric acid in the atmosphere can attack mortar to form lime sulfate, which may be absorbed by brick, crystallize there, and cause brick disintegration. Some brick and some mortar, neither of which alone cause serious efflorescence, can react to produce large amounts of efflorescence [*218*]. Sulfate in brick, which is free of efflorescence, may react with low-alkali cement (less than 0.21% free alkali) in mortar to produce water-soluble alkali sulfate salts and copious efflorescence [*203*]. This interaction, called the Camerman effect, was challenged by Butterworth [*150*].

Florescence is a physical effect, involving the deposition of salt in solid form. Sulfate attack is a chemical effect, which may occur when masonry remains wet and sulfates in brick are transferred to mortar where they react with tri-calcium aluminate to form expansive ettringite [*45*]. Mortar expansion attributable to sulfates from brick can be as much as 0.023% [*224,225*].

Efflorescence should not be confused with scumming, which is an insoluble stain, usually white but may be pink or yellow, formed on the surface of bricks during the manufacturing process. It is often calcium or magnesium aluminosilicate, formed from magnesium or calcium sulfate impurities in the raw materials [*201*]. There is no evidence that scum affects durability. Scum, if present, can be observed when the bricks come from the kiln.

Jackson provides a descriptive bibliography on scumming and efflorescence with 232 entries from 1877 through 1924 [*212*]. Brownell [*199*] provided in 1949 a discussion of the factors influencing efflorescence in bricks and a bibliography with 107 entries.

Efflorescence appears, if at all, only after exposure to water. Under suitable conditions efflorescence may be derived entirely from the action of the atmosphere [*217*]. The way to prevent or reduce efflorescence is to reduce salt content of materials and prevent water permeance [*210*].

Environment

"The service conditions of a brick which govern its wetting and freezing, and thus its durability, vary greatly depending on geographical location of the particular building, local weather conditions, the orientation of the wall in which the brick serves, and its location in that wall" [*87,140*]. The significant aspects of weather which affect florescence are rain, acid rain, temperature, humidity, and wind. Wind-driven rain affects water permeance. Sulfuric acid rain can react with calcium, magnesium, and alkali to form sulfates. Chloride florescence may occur near seawater [*135,194*].

Weather conditions favorable to efflorescence [*173*] are: (1) An initial soaking rain of at least one inch (25.4 mm) in 24 hours, spread over several hours. Rain penetration into masonry is aided by wind; (2) Drying at air temperatures of 35°F (2°C) to 50°F (10°C) for two days or one day with a breeze; and (3) Repeated wet-dry cycles.

Slow evaporation of water from the brick surface favors efflorescence, and rapid drying favors cryptoflorescence [*208*]. Florescence is greatest when evaporation is greatest; i.e., high wind and high vapor pressure differential. The evaporation rate depends also on brick porosity; i.e., the size, number, shape, and arrangement of pores [*208*]. The mechanics of evaporation from bricks were described by Laurie and Milne [*215*].

The intensity of wetting of a wall due to rain is much greater near the top of the wall than at its base, which explains why decayed brick are more prevalent near the tops of wall, since frost damage is related to exposure.

The temperature at the center of a brick on a south wall may be 9°C (15°F) above freezing while the exterior air temperature is 9°C (15°F) below freezing [*140*]. The surface temperature of a south-facing brick wall may rise from well below freezing to nearly 70°F (21°C) while that of a north-facing wall may rise only a few degrees above air temperature and stay well below freezing. The variables which control the temperature of masonry and, therefore, its freezing frequency are discussed by Grimm [*156*] and are listed in Appendix IV.

The number of annual freeze-thaw cycles at the center of a brick in a south wall may be 70% greater than in a north wall. The number of freeze-thaw cycles is greater near the exterior face of the wall than at its center and lesser near the interior face of the wall. Porous brick, being less responsive to heat input, undergoes fewer freeze-thaw cycles. Moisture content during winter varies greatly with location, orientation, and absorption by masonry materials.

A weathering index for brick was developed by Grimm in 1956 [*14*]. The index for any locality is the product of the average annual number of freezing cycle days and the average annual winter rainfall in inches. The index has been incorporated into specifications of the American Society for Testing and Materials, ASTM C 33, C 216, and C 62.

Brick masonry in contact with earth has a much more severe exposure than it does in walls above ground, because the masonry is more apt to contain more salt when dry and water when frozen. Thus brick masonry in retaining walls, foundation walls, and planter boxes in a severe climate is less durable than the same masonry in a wall enclosing heated space [*152*].

Similarly, brick masonry in the exterior wythe of an insulated wall has a more severe exposure than in an uninsulated solid wall. It is more apt to be frozen at the lower temperatures prevailing in the exterior wythe of the insulated cavity wall.

Brick masonry in a nearly horizontal position has a much more severe exposure than in a vertical position, because it is more apt to absorb more water. Brick masonry in pavements, copings, and sills in a severe climate is less durable than the same masonry in a wall.

Brick masonry in walls exposed to weather on both sides has a much more severe exposure than in walls enclosing heated space, because it is more apt to freeze more often. Brick masonry in parapets, chimneys, wing walls, and fences in a severe climate is less durable than the same masonry in a wall enclosing heated space.

Architectural Engineering

Poor masonry design contributes to masonry disintegration [*111,132, 139,173,210*].

The use of coping to prevent water entry at the top of the wall and through-wall flashing at the base of the wall to prevent upward migration of groundwater is necessary to masonry durability [*75*]. Sills and copings should project beyond the face of the wall and be provided with drips [*139*]. Kamp [*132*] recommended water stops in sealant joints.

Air leakage outward at the top of a building, especially at windows, can cause excessive condensation. Covering the back of parapets with roof membranes and flashing causes ice lensing. Condensation from exfiltration in glazed brick masonry without a vapor barrier has caused disintegration. Ventilation of cavity walls is important [*123*]. McBurney [*134*] recommended a vapor barrier in glazed brick walls. A vapor barrier would also reduce florescence.

All of those factors, which affect water permeance of masonry, also affect its durability [*114*]. Grimm provides a review of the literature on water permeance of masonry walls [*127*], and offers design recommendations [*128*]. The subject of water permeance is outlined in Appendix II.

If florescence is perceived as a problem, the designer should require brick rated as "no efflorescence" under ASTM C 67-81, and should specify low-alkali portland cement under ASTM Specification for Portland Cement (C 150-83a) and nonstaining masonry cement under ASTM Specification for Masonry Cement (C 91-83) [*214*]. The use of a dispersing agent in mortar to reduce water content will also reduce efflorescence [*216*], but its effect, if any, on bond should be considered.

Brick Specifications

Durability is improved by requiring that bricks conform to the requirements for Grade SW under the several ASTM brick specifications. Quality control in brick manufacturing is beyond the scope of this paper. However, firing temperature is well known to affect durability of brick [*158*]. In 1936 Butterworth [*28*] admitted that there was "no single test for discriminating between well-fired and underfired brick." However, a study of X-ray diffractographs of fired samples of brick clay can indicate the temperature at which brick have been fired with a standard deviation of 14°F (8°C).

Although laminations in brick may be expected to affect brick strength and durability, no study appears to have been made specifically on a relationship between laminations and brick performance [*74*]. The presence and pattern of brick coring has no significant affect on resistance of brick masonry to rain penetration [*18, 145*].

Sand-finish brick may have sand on the mortar bonding surface which reduces bond strength, increases water permeance, and contributes to efflorescence [*149*]. Unbroken normal die skin may induce frost damage [*78*]. Dense impervious brick which absorb water very slowly can be bonded satisfactorily with mortar [*66*].

Robinson et al [*76*] in 1977 analyzed freeze-thaw test data on 5217 bricks from 34 plants, which produce about a quarter of all U.S. bricks. Those authors proposed new specification requirements to discriminate between durable and nondurable brick. Their durability index was expressed as follows, where all properties are determined in accordance with ASTM C 67:

$$Y = IRA\ 0.10[1 - (C/B)]^{-1} + C - f_b - 4$$

where

Y = durability index, dimensionless,
IRA = initial rate of water absorption, g/min (30 in.^{2}),
C/B = ratio of 24-h cold water absorption to 5-h boiling water absorption, dimensionless,
C = 24-h cold water absorption, %, and
f_b = compressive strength of brick, ksi.

Based on that equation, Robinson and his co-workers proposed specification requirements which would accept 15% of the bad brick and reject 18% of the good brick, bad and good being differentiated by ability to pass the ASTM freeze-thaw test. Based on the same test data and equation, Grimm proposes a maximum durability index of 10.85, which would accept 7.2% of the bad brick and reject an equal percentage of the good brick.

Construction

Poor masonry construction reduces durability [*210*]. All of those aspects of workmanship which affect water permeance also affect durability. At least ten authors have emphasized that workmanship affects water permeance more than any other factor. In a review of the literature on water permeance, Grimm [*127*] cites the following workmanship considerations: (1) use of proper mortar materials, their proportioning, and mixing; (2) proper selection, preparation, and culling of brick, and their wetting if necessary; (3) the complete solidity of all mortar joints; i.e., head, bed, and collar joints; (4) cleanliness of cavities; (5) concave or "V" tooling of the exterior face of mortar joints; (6) proper placement and sealing of flashing with end dams and weep holes, and (7) proper materials and construction of sealant joints.

Brick and mortar should be selected with due regard for their compatibility. Although air entrainment improves mortar durability, excessive air entrainment (above 15%) reduces bond strength, increases water permeance, and to that extent reduces durability.

Grimm and Halsell [*155*] discuss the several aspects of quality control in brick masonry construction.

Maintenance

The law in Chicago and New York requires at intervals of five to ten years the inspection of building facades by an architect or engineer. Prudence would suggest that a property manager should require close-up visual inspections of masonry at three-year intervals by a less competent person, perhaps by a well instructed window washer or other person working on light, suspended scaffolding.

The inspection should include visual observation and location of defective sealants in the construction joints around wall openings and in horizontal and vertical expansion joints in masonry. The average life of a sealant is seven years [*116*], and the water permeance of buildings is often left entirely to the integrity of such joints. The inspection should also include the location, size, and extent of all stains, cracks, spalls, or other evidence of masonry distress.

Poor maintenance reduces the durability of masonry [*210*]. Grimm [*125*] provides a guide to the literature on masonry maintenance and restoration with 120 citations. The topics covered include aesthetics, masonry units, mortar, flashing, metals in masonry, repair, cleaning, and coatings.

In another paper Grimm [*126*] discusses 15 types of coatings for masonry; i.e., six types of paint, three types of membranes, three plasters, two stuccos, and surface grouting. Coatings should not be applied to masonry indiscriminantly. Colorless coatings are not durable, often ineffective, and can cause masonry disintegration. It is not wise to use coatings to control efflorescence [*110,123,164,194,203,223*]. They can retard the normal drying process and increase the probability of the masonry having high water content when frozen.

Conclusions

The ASTM freeze-thaw test for bricks with all its imperfections is the best method presently known to differentiate between durable and nondurable brick. The specification of a maximum durability index of 10.85 would probably provide adequate assurance of durability and a reasonable balance between consumer and producer interests. No brick specification requirement would provide durable brick masonry without prudent architectural engineering specifications for durable and compatible mortar and reduced water permeance of masonry [*210*]. Good materials and design will not result in durable brick masonry without good construction workmanship and proper maintenance. The manufacturers of brick, lime, cement, and sand, the architect and engineer, the contractor, and the building owner must share responsibility for the durability of brick masonry. Keeping the masonry as dry as possible is the single most important variable in masonry durability.

APPENDIX I

Subject Outline: Durability of Masonry

1. Distructive Agents
    - 1.1 Climatological
        - 1.1.1 Water
        - 1.1.2 Heat
        - 1.1.3 Air
    - 1.2 Geological
    - 1.3 Biological
        - 1.3.1 Bacteria
        - 1.3.2 Plants
        - 1.3.3 Animals
    - 1.4 Chemical
        - 1.4.1 Constituents
        - 1.4.2 Contaminants
        - 1.4.3 Pollutants

2. Mechanics of Destruction
    2.1 Crystallization
        2.1.1 Water
        2.1.2 Salt
    2.2 Hydration
    2.3 Oxidation
    2.4 Osmosis
    2.5 Deformation
        2.5.1 Thermal
        2.5.1 Moisture
        2.5.3 Wet-dry
        2.5.4 Freezing
        2.5.5 Elastic
        2.5.6 Creep
        2.5.7 Shrinkage
        2.5.8 Differential movement
3. Materials: Manufacturers and Distributors
    3.1 Physical Properties of Materials
        3.1.1 Porosity
        3.1.2 Strength
        3.1.3 Test methods
        3.1.4 Specifications
    3.2 Product Quality Control
4. Design: Architects and Engineers
    4.1 Exposure Severity
        4.1.1 Climate and environment
        4.1.2 Position of structure
        4.1.3 Position in structure
    4.2 Water Permeance Control (See Appendix II)
    4.3 Crack Control (See Appendix III)
5. Construction: Contractors and Inspectors
    5.1 Construction Quality Control
        5.1.1 Water permeance (see Appendix II)
        5.1.2 Expansion joints (see Appendix III)
        5.1.3 Wall anchorage (see Appendix III)
6. Maintenance: Owner
    6.1 Periodic Inspection
        6.1.1 Sealant joints
        6.1.2 Stains
        6.1.3 Leaks
        6.1.4 Cracks
    6.2 Maintenance Techniques
        6.2.1 Water permeance control (see Appendix II)
        6.2.2 Crack control (see Appendix III)

APPENDIX II

Subject Outline: Water Permeance of Masonry Walls

1. Consequences of Water Permeance
    1.1 Freeze-thaw disintegration

- 1.2 Florescence
  - 1.2.1 Efflorescence
  - 1.2.1 Disentegration due to cryptoflorescence
- 1.3 Metal corrosion
- 1.4 Wood decay
- 1.5 Increased fuel costs
- 1.6 Deterioration of:
  - 1.6.1 Interior finish
  - 1.6.2 Building contents
- 1.7 People problems
  - 1.7.1 Tenant inconvenience
  - 1.7.2 Owner anxiety

2. Design: Architects and Engineers
   - 2.1 Climatology versus Wall Type
     - 2.1.1 Driving rain severity
     - 2.1.2 Selection of wall type
   - 2.2 Brick-Mortar Compatibility
     - 2.2.1 Brick suction
     - 2.2.2 Water retentivity of mortar
     - 2.2.3 Construction temperature
   - 2.3 Flashing
     - 2.3.1 Location and configuration
     - 2.3.2 Seals and dams
   - 2.4 Crack Control (See Appendix III)
     - 2.4.1 Materials quality
     - 2.4.2 Mortar joint configuration
     - 2.4.3 Sealant joints
       - 2.4.3.1 Location and size
       - 2.4.3.2 Configuration
     - 2.4.4 Wall anchorage
     - 2.4.5 Masonry misuse
       - 2.4.5.1 Copings and sills
     - 2.4.6 Masonry strength
   - 2.5 Specifications
     - 2.5.1 Vapor barriers
     - 2.5.2 Weep holes
     - 2.5.3 Cavity wall vents
     - 2.5.4 Metal copings
3. Materials: Manufacturers and Distributors
   - 3.1 Cracked masonry units
   - 3.2 Clean masonry units
   - 3.3 Physical properties of materials
4. Construction: Contractor and Inspector
   - 4.1 Culled materials
   - 4.2 Well proportioned and mixed mortar
   - 4.3 Units placed in plastic mortar
   - 4.4 Well filled mortar joints
   - 4.5 Clean cavities and sealant joints
   - 4.6 Properly installed flashing
   - 4.7 Well tooled mortar joints
   - 4.8 Well sealed (caulked) joints
   - 4.9 Weather adaptation

5. Maintenance: Owner
   5.1 Periodic Inspection
      5.1.1 Sealant joints
      5.1.2 Stains
      5.1.3 Weathering
      5.1.4 Cracks
   5.2 Techniques
      5.2.1 Sealant replacement
      5.2.2 Cleaning masonry
      5.2.3 Surface grouting
      5.2.4 Tuck pointing
      5.2.5 Vegetation removal
      5.2.6 Coatings
      5.2.7 Crack repair (see Appendix III)
   5.3 Alternatives
      5.3.1 Demolition
      5.3.2 Reconstruction

APPENDIX III

Subject Outline: Cracked Masonry

1. Objections to Cracks
   1.1 Aesthetic and Psychological
   1.2 Water Permeance
   1.3 Structural Stability
2. Crack Classification
   2.1 Causality
      2.1.1 Facial separation cracks
      2.1.2 Cracks in masonry units and in mortar
         2.1.2.1 Manufacturing defects
         2.1.2.2 Weathering
         2.1.2.3 Strain
      2.1.3 Structural cracks
         2.1.3.1 Compression
         2.1.3.2 Tension
         2.1.3.3 Shear
   2.2 Size
      2.1.1 Negligible (0 to 0.1 mm)
      2.2.2 Very slight (0.1 to 1 mm)
      2.2.3 Slight (1 to 5 mm)
      2.2.4 Moderate (5 to 15 mm)
      2.2.5 Severe (15 to 25 mm)
      2.2.6 Very Severe (greater than 25 mm)
3. Crack Appearance
   3.1 Pattern
      3.1.1 Vertical split
      3.1.2 Horizontal separation
      3.1.3 Diagonal step
      3.1.4 Cogged

3.2 Visibility
3.2.1 Size versus distance (distance < 3440 width)
3.2.2 Exposure time, contrast, and illumination
4. Cause of Structural Cracks
4.1 Foundation Displacement
4.1.1 Moisture change in plastic soil
4.1.2 Soil subsidence
4.1.3 Uneven settlement
4.2 Structural Frame Movement
4.2.1 Elastic deformation
4.2.2 Thermal displacement
4.2.3 Concrete creep and shrinkage
4.2.4 Side sway (drift)
4.2.5 Beam deflection
4.2.6 Metal corrosion
4.2.7 Dimensional change in wood
4.2.8 Shelf angle rotation and deflection
4.3 Masonry Movements
4.3.1 Temporary moisure expansion
4.3.2 Permanent moisture expansion
4.3.3 Thermal displacement
4.3.4 CMU shrinkage
4.3.5 Mortar shrinkage
4.3.6 Freezing expansion
4.3.7 Bending deflection
4.3.8 Compression buckling
5. Design: Architects and Engineers
5.1 Flexible Anchorage
5.2 Expansion/Contraction Joints
5.2.1 Size
5.2.2 Location
5.3 Proper Use of Materials
6. Materials: Manufacturers and Distributors
6.1 Cracked Masonry Units
6.2 Physical Properties of Materials
7. Construction: Contractors and Inspectors
7.1 Culled Material
7.2 Clean and Well Sealed Expansion Joints
7.3 Anchor Type, Size, and Location
7.4 Well Tooled Joints
7.5 Well Filled Joints
7.6 Plain and Straight
7.7 Well Proportioned and Mixed Mortar
7.8 Units Placed in Plastic Mortar
7.9 Weather Adaptation
7.10 Proper Positioning
8. Maintenance: Owner
8.1 Periodic Inspection
8.1.1 Sealant joints
8.1.2 Cracks

- 8.1.3 Masonry displacement
  - 8.1.3.1 In plane
  - 8.1.3.2 Out of plane
- 8.2 Techniques
  - 8.2.1 Sealant joint replacement
  - 8.2.2 Tuck pointing
  - 8.2.3 Surface grouting
  - 8.2.4 Epoxy injection

APPENDIX IV

Factors Affecting the Mean Temperature of Masonry

1. Climatological
   - 1.1 Air temperature
     - 1.1.1 Exterior
     - 1.1.2 Interior
   - 1.2 Wind velosity
   - 1.3 Atmospheric clarity
   - 1.4 Moisture content of masonry
2. Thermal Properties of Materials
   - 2.1 Conductivity
   - 2.2 Specific heat
   - 2.3 Density
   - 2.4 Thickness
   - 2.5 Overall heat transmission
   - 2.6 Radiant emissivity
   - 2.7 Radiant reflectance of surroundings
3. Architectural Design
   - 3.1 Tilt angle of masonry
   - 3.2 Tilt angle of surroundings
   - 3.3 Wall azimuth
4. Solar Angles
   - 4.1 Date and hour
   - 4.2 Longitude and latitude

Bibliography

Brick

[*1*] "Absorption of Clay Brick," Technical News Bulletin 141, U.S. National Bureau of Standards, Washington, DC, 1929, pp. 3–4.

[*2*] "The Brick Cemetery at Watford," *British Clayworker*, Vol. 42, No. 499, 1933, pp. 231–232.

[*3*] "Pore Structure and Weather Resistance of Structural Clay Products," Technical News Bulletin 223, U.S. National Bureau of Standards, Washington, DC, 1935, p. 116.

[*4*] "Report of ASTM Committee C-3 on Brick Including Report of Subcommittee XI on Weathering and Porosity," *ASTM Proceedings*, American Society for Testing and Materials, Philadelphia, Vol. 35, 1935, pp. 251–255.

[*5*] "Weathering of Clay Brick," Technical News Bulletin 216, U.S. National Bureau of Standards, Washington, DC, 1935, p. 41.

[6] "Resistance of Brick to Freezing and Thawing," Technical News Bulletin 217, U.S. National Bureau of Standards, Washington, DC, 1935, pp. 51-52.
[7] "Effect of Repeated Freezing and Thawing upon the Pore Structure of Brick," Technical News Bulletin 225, U.S. National Bureau of Standards, Washington, DC, 1936, p. 7.
[8] "Influence of Repeated Freezing and Thawing on the Relation of Water Absorbed to Pore Volume of Bricks," Technical News Bulletin 234, U.S. National Bureau of Standards, Washington, DC, 1936, pp. 86-87.
[9] "Relation of Pore Size of Bricks to Their Resistance to Disintegration by Freezing," Technical News Bulletin 243, U.S. National Bureau of Standards, Washington, DC, 1937, p. 73.
[10] "Clay Salinity and Efflorescence," *The British Clayworker*, Vol. 46, No. 545, Sept. 1937, p. 194.
[11] "Weathering Properties of Building Bricks," Technical Information on Building Materials, TIBM57, U.S. National Bureau of Standards, Washington, DC, Dec. 1937, p. 4.
[12] "Experiments on the Weathering Properties of Bricks," *The British Clayworker*, Vol. 55, No. 651, 1947, pp. 51-52.
[13] "Spalling of Glazed Bricks Pointed in Keene's Cement," *The British Clayworker*, Vol. 56, No. 666, 1947, pp. 165-166.
[14] "Derivation of Weathering Index for Bricks," ASTM Bulletin No. 217, American Society for Testing and Materials, Philadelphia, 1956, pp. 39-40.
[15] "Sulphate Attack on Brickwork," *Building Research Establishment Digest*, No. 89, Building Research Station, Garston, Watford, England, 1972.
[16] "Testing Methods for Natural and Artificial Stones. I—Natural Stones. II—Baked Clay Masonry Units," *Materials and Construction*, RILEM, Paris, Vol. 5, No. 28, July/Aug. 1972, pp. 247-257.
[17] *Specifications for Clay Bricks and Blocks*, BS 3921, 1974, British Standards Institution, London, England, May 1974, p. 10.
[18] "Tests Defining the Structure," *Materials and Construction*, RILEM, Paris, Vol. 13, No. 79, May/June 1980, p. 175.
[19] "Perforated Clay Brick," *Building Research Establishment Digest*, No. 273, Her Majesty's Stationery Office, London, May 1983.
[20] Astbury, N. F. and Vyse, J., "A New Method for the Study of Pore Size Distribution," *Transactions of the British Ceramic Society*, Stoke-on-Trent, England, Vol. 70, No. 3, 1971, pp. 77-85.
[21] Baldwin, L. W., "Frost Resistence—Discussion," *Journal of the British Ceramic Society*, Stoke-on-Trent, England, Vol. 3, No. 1, 1966, pp. 145-149.
[22] Bauleke, M. P., "Effects of Chemicals on Colour and Durability of Iowa Devonian Shale Products," *Journal of Applied Chemistry*, Vol. 7, No. 12, 1957, pp. ii-515.
[23] Bloor, J. W., "The Saturation Freezing Test," *Journal of the British Ceramic Society*, Stoke-on-Trent, England, Vol. 1, No. 2, 1964, pp. 226-228.
[24] Bonnell, D. R. G., "Some Problems Connected with Porous Building Materials," *Chemistry and Industry*, Vol. 57, No. 9, 1938, pp. 195-198.
[25] Booth, C. A., "Strength and Absorption Tests for Clay and Shale Bricks," *Canadian Ceramic Society Journal*, Toronto, Ontario, Vol. I, No. 1, 1932, pp. 54-57.
[26] Brownell, W. E., *Structural Clay Products*, Springer-Verlag, Wien, New York, 1976.
[27] Butterworth, B., "The Correlation of Laboratory Tests with the Weathering Properties of Bricks," *Transactions of the British Ceramic Society*, Stoke-on-Trent, England, Vol. 33, No. 11, 1934, pp. 495-526; *British Clayworker*, Vol. 43, No. 513, 1935, pp. 185-194.
[28] Butterworth, B., "Frost Tests on Bricks and Tiles and Their Limitations," International Association for Testing Materials, London Congress, 1937, pp. 383-384; *Claycraft*, Vol. 10, No. 9, 1937, p. 461.
[29] Butterworth, B., "XXII. Some Effects of Soluable Salts in Clay Products," *Transactions of the British Ceramic Society*, Stoke-on-Trent, England, Vol. 36, No. 5, 1937, pp. 233-242.
[30] Butterworth, B., "The Absorption of Water by Clay Building Bricks and Related Properties—The Rate of Absorption of Water by Part by Saturated Bricks," *Transactions of the British Ceramic Society*, Vol. 45 and 46, May 1946, March-April 1947.
[31] Butterworth, B., "Frost Resistance of Fired Clay—A Perennial Problem," *British Clayworker*, Vol. 61, No. 722, 1952, pp. 85-89; No. 723, pp. 122-124; *The Brick Builder*, Sept. 1952.

[32] Butterworth, B., "The Practical Significance of Tests for Soluble Salts in Clay Products," *Transactions of the British Ceramic Society*, Stoke-on-Trent, England, Vol. 52, No. 4, 1953, pp. 174–178.

[33] Butterworth, B., "The Properties of Clay Building Materials," *Ceramics—A Symposium*, British Ceramic Society, Stoke-on-Trent, England, 1953, pp. 824–877.

[34] Butterworth, B., "The Work of the T.B.E. (Tuiles et Briques Europeenes) Frost Committee for the Brick Development Association Ltd.," *Claycraft*, Vol. 34, No. 11, Aug. 1961, pp. 346–348.

[35] Butterworth, B., "The Frost Resistance of Bricks and Tiles: A Review," *Journal of the British Ceramic Society*, Stoke-on-Trent, England, Vol. 1, No. 2, 1964, pp. 203–223.

[36] Butterworth, B., "The Indirect Appraisal of [Brick] Durability. Part I," *Transactions of the British Ceramic Society*, Stoke-on-Trent, England, Vol. 63, No. 11, 1964, pp. 639–646.

[37] Butterworth, B., "The Indirect Appraisal of [Brick] Durability. Part II," *Transactions of the British Ceramic Society*, Stoke-on-Trent, England, Vol. 63, No. 11, 1964, pp. 647–661.

[38] Butterworth, B., "Laboratory Tests and the Durability of Bricks, Part II. Recording Comparison and Use of Outdoor Exposure Tests," *Transactions of the British Ceramic Society*, Stoke-on-Trent, England, Vol. 63, Nov. 1964, pp. 615–628.

[39] Butterworth, B., "Laboratory Tests and the Durability of Bricks. IV. The Indirect Appraisal of Durability," *Transactions of the British Ceramic Society*, Stoke-on-Trent, England, Vol. 63, No. 11, 1964, pp. 639–646.

[40] Butterworth, B. and Baldwin, L. W., "Laboratory Tests and the Durability of Bricks. V. The Indirect Appraisal of Durability" (continued), *Transactions of the British Ceramic Society*, Stoke-on-Trent, England, Vol. 63, No. 11, 1964, pp. 647–661.

[41] Butterworth, B., "Testing the Frost Resistance of Bricks," *Transactions of the Ninth International Ceramic Congress*, Brussels, Ref. DEC: E31/D81, 1964, pp. 439–458.

[42] Butterworth, B., "The Effect of Weathering Upon Rate of Absorption," *Transactions of the British Ceramic Society*, Stoke-on-Trent, England, Vol. 65, No. 1, 1966, pp. 51–57.

[43] Butterworth, B. and Carter, E. F., "Laboratory Tests and Durability of Bricks. Part VIII. Frost Dilatometry—Modified Methods," *Transactions of the British Ceramic Society*, Stoke-on-Trent, England, Vol. 66, No. 1, 1967, pp. 1–12.

[44] Butterworth, B., "Ceramic Building Materials, Durability, *Claycraft*, London, England, Vol. 41, No. 10, July 1968, pp. 332–335.

[45] Butterworth, B., "Ceramic Building Materials," *Claycraft*, London, England, Vol. 41, No. 11, Aug. 1968, pp. 381–384.

[46] Butterworth,, B. and Baldwin, L. W., "Durability of Bricks Made with Pulverised Fuel Ash," *Proceedings of a Joint Meeting of the Building Materials Section of the British Ceramic Society and the Institute of Clay Technology*, 12–13, April 1972, Stoke-on-Trent, England, 1972, pp. 21–25.

[47] Butterworth, B., Baldwin, L. W., and Newman, A. J., "Changes in Strength of Buried Bricks," *Journal of the British Ceramic Society*, Stoke-on-Trent, England, Vol. 77, 1978.

[48] Carlson, Orvar, "The Influence of Submicroscopic Pores on the Resistance of Bricks Towards Frost" (In English), *Transactions of Chalmers University of Technology*, Gothenburg, Sweden, No. 212, 1959, p. 13.

[49] Clews, F. H., "Frost Action and Elastic Behaviour of Clay Products," *Journal of the British Ceramic Society*, Stoke-on-Trent, England, Vol. 1, No. 2, 1964, pp. 228–229.

[50] Cobb, J. W., "Note on Disintegration by Crystallization and Freezing," *Transactions of the American Ceramic Society*, Columbus, OH, Vol. 11, 1909.

[51] Davenport, S. T. E., "The Properties of Bricks," *Claycraft*, London, England, Vol. 41, No. 12, 1968, pp. 400–404.

[52] Davison, J. I., "Outdoor Freeze-Thaw Cycling of Dry-Press Bricks," Internal Report No. 308, National Research Council of Canada, Division of Building Research, Ottawa, Canada, 1964.

[53] Davison, J. I., "Durability Studies on Brick Used in the Atlantic Provinces," *Journal of the Canadian Ceramic Society*, Vol. 44, 1975, pp. 23–29; National Research Council of Canada, Division of Building Research, DBR Paper No. 684, NRC 15365, 1975.

[54] Davison, J. I., "Effect of Temperature on Brick Suction," *Journal of Testing and Evaluation*, Vol. 10, No. 3, May 1982, pp. 81–82.

[55] Davison, J. I., "Linear Expansion Due to Freezing and Other Properties of Bricks," *Pro-*

ceedings of The Second Canadian Masonry Symposium, Carleton University, Ottawa, Ontario, Canada, June 1980, p. 13-24.

[*56*] Douty, D. E. and Beebe, L. L., "Some Further Experiments Upon the Absorption, Porosity and Specific Gravity of Building Brick," *ASTM Proceedings*, American Society for Testing and Materials, Philadelphia, Vol. XI, 1911, p. 767.

[*57*] Foster, H. D., "Weathering Test Procedures for Clay Products," *ASTM Proceedings*, American Society for Testing and Materials, Vol. 31, Part II, 1931.

[*58*] Furlong, Irving, "Alkali Attack on Concrete Roads and Building Brick," *Engineering News Record*, New York, Vol. 89, July 22, 1922, pp. 64-67.

[*59*] Grattan-Bellew, P. E., and Litvan, G. G., "X-Ray Diffraction Method for Determining the Firing Temperature of Clay Brick," *American Ceramic Society Bulletin*, American Ceramic Society, Columbus, OH, Vol. 57, No. 5, May 1978, pp. 493-495.

[*60*] Grimm, C. T. and Houston, J. T., "Structural Significance of Brick Water Absorption," *Masonry: Past and Present, ASTM STP 589*, American Society for Testing and Materials, Philadelphia, 9175, pp. 272-289.

[*61*] Gummerson, R. J., et al: "The Suction Rate and the Sorptivity of Brick," *Journal of The British Ceramic Society*, Stoke-on-Trent, England, Vol. 80, 1981, p. 150-152.

[*62*] Haynes, J. M., "Frost Action as a Capillary Effect," *Claycraft*, Vol. 37, No. 10, 1964.

[*63*] Johnson, P. V. and Plummer, H. C., "Some Factors Affecting Durability of Structural Clay Products Masonry," *Symposium on Some Approaches to Durability in Structures, ASTM STP 236*, American Society for Testing and Materials, Philadelphia, 1958, pp. 3-13.

[*64*] Jones, J. C., "The Relation of Hardness of Brick to Their Resistance to Frost," *Transactions of the American Ceramic Society*, Columbus, OH, Vol. 9, 1907, p. 528.

[*65*] Lewis, R. C., "A Relationship Between the Percent Absorption, The Modules of Rupture, and the Electrical Conductivity of Building Brick," *Journal of the American Ceramic Society*, Columbus, OH, Vol. 15, No. 10, 1932, pp. 574-581.

[*66*] Litvan, G. G., "Testing the Frost Susceptibility of Bricks," *Masonry: Past and Present*, Symposium, ASTM STP 589, American Society for Testing and Materials, Philadelphia, June 1974, p. 123.

[*67*] Llewellyn, H. M. and Butterworth, B., "Laboratory Test and the Durability of Bricks—Part III: Some Conventional Laboratory Freezing Tests," *Transactions of the British Ceramic Society*, Stoke-on-Trent, England, Vol. 63, No. 11, 1964, pp. 629-637.

[*68*] McBurney, J. W., "The Water Absorption and Penetrability of Brick," *ASTM Proceedings*, American Society for Testing and Materials, Philadelphia, Vol. 29, Part II, 1929, pp. 711-739.

[*69*] McBurney, J. W., "The Weathering of Structural Clay Products—A Review," *ASTM Proceedings*, American Society for Testing and Materials, Philadelphia, Vol. 31, Part II, 1931, p. 745.

[*70*] McBurney, J. W. and Lovewell, C. E., "Strength, Water Absorption and Weather Resistance of Building Brick Produced in the United States," *ASTM Proceedings*, American Society for Testing and Materials, Philadelphia, Vol. 33, Part II, 1933, p. 636.

[*71*] McBurney, J. W., "The Relation of Freezing and Thawing Resistance to Physical Properties of Clay and Shale Building Brick," *ASTM Proceedings*, American Society for Testing and Materials, Philadelphia, Vol. 35, Part 1, Appendix I, 1935, p. 247.

[*72*] McBurney, J. W., "Water Absorption of Building Bricks," *ASTM Proceedings*, American Society for Testing and Materials, Philadelphia, Vol. 36, Part I, 1936, pp. 260-271.

[*73*] McBurney, J. W., "Relations Between Results of Laboratory Freezing and Thawing and Several Physical Properties of Certain Soft-Mud Bricks," *ASTM Proceedings*, American Society for Testing and Materials, Philadelphia, Vol. 42, 1942, p. 837.

[*74*] McBurney, J. W., Richmond, J. C., and Copeland, M. A., "Relations Among Certain Specification Properties of Building Brick and Effects of Differences in Raw Materials and Methods of Forming," *Journal of the American Ceramic Society*, Columbus, OH, Vol. 35, No. 12, 1952, pp. 309-318.

[*75*] McBurney, J. W. and Johnson, P., "Durability of De-Aired Bricks," *Journal of the American Ceramic Society*, Columbus, OH, Vol. 39, No. 5, May 1956, pp. 159-168.

[*76*] McHugh, G. P. and Knight, E. L., "The Effect of Dissolution on the Durability of Fletton Bricks," *Chemistry and Industry*, Vol. 51, No. 14, 1932, pp. 107-110.

[*77*] McIntyre, W. A., "Factors Governing the Durability of Clay Building Materials," *British*

Clayworker, Vol. 37, No. 436, 1928, pp. 195-205; *Transactions of the British Ceramic Society*, Stoke-on-Trent, England, Vol. 28, No. 3, 1929, pp. 101-123.

[*78*] Orton, E., Jr., "A Comparison Between the Absorption, Crushing Strength and Resistance to Artificial Freezing of Some Ohio Building Bricks," *Transactions of the American Ceramic Society*, Columbus, OH, Vol. 18, 1916, p. 686.

[*79*] Palmer, L. A. and Hall, J. V., "Some Results of Freezing and Thawing Tests Made with Clay Brick," *ASTM Proceedings*, American Society for Testing and Materials, Philadelphia, Vol. 30, Part II, 1930, p. 767.

[*80*] Parsons, D. E., "Bibliography on the Weathering of Structural Clay Products," *ASTM Proceedings*, American Society for Testing and Materials, Philadelphia, Vol. 31, Part II, June 1931, pp. 825-834.

[*81*] Parsons, D. E., "Comparison of Natural Weathering with Laboratory Tests of Clay Brick," *ASTM Proceedings*, American Society for Testing and Materials, Philadelphia, Vol. 35, Part I, Appendix II, 1935, p. 252.

[*82*] Peake, F. and Ford, R. W., "A Comparison of The Vacuum and Boiling Methods for Measuring the Water Absorption of Bricks," *Transactions of the British Ceramic Society*, Stoke-on-Trent, England, 81, 1982, pp. 160-162.

[*83*] Phillips, J. G., *The Physical Properties of Canadian Building Bricks*, Department of Mines and Resources, Ottawa, Canada, 1947.

[*84*] Richmond, J. C. and McBurney, J. W., "Effect of Freezing Temperature in Freezing and Thawing Tests of Brick," *ASTM Proceedings*, American Society for Testing and Materials, Philadelphia, Vol. 41, 1941, pp. 967-974.

[*85*] Ritchie, T. and Meincke, H. R., *Capillary Absorption of Some Canadian Building Bricks*, Research Paper No. 8, Div. of Building Research, National Research Council, Ottawa, Canada, April 1953.

[*86*] Ritchie, T., "Factors Affecting Frost Damage to Clay Bricks," Building Research Note, National Research Council of Canada, Division of Building Research, Ottawa, Canada, Note No. 62, 1968.

[*87*] Ritchie, T., "Freeze-Thaw Action on Brick," *Journal of the Canadian Ceramic Society*, Vol. 41, 1972, pp. 1-6; reprinted by National Research Council of Canada, Division of Building Research, Research Paper No. 55, NRC 13136, Ottawa, Canada, 1972.

[*88*] Ritchie, T., "The Test Method of Freezing Bricks: Its Influence on Their Durability," *Journal of the Canadian Ceramic Society*, Vol. 44, 1975, pp. 21-22; National Research Council of Canada, Division of Building Research, DBR Paper No. 683, NRCC 15361, Ottawa, Canada.

[*89*] Ritchie, T., Measurement of Laminations in Brick," *American Ceramic Society Bulletin*, Vol. 54, No. 9, 1975, pp. 725-726.

[*90*] Robinson, G. S., *An Accelerated Test Method for Predicting the Durability of Brick*, Masters Thesis, Clemson University, Clemson, SC, Sept. 1976.

[*91*] Robinson, G. C., Holman, J. R., and Edwards, J. F., "Relation Between Physical Properties and Durability of Commercially Marketed Brick," *Ceramic Bulletin*, American Ceramic Society, Columbus, OH, Vol. 36, No. 12, 1977, pp. 1071-1076.

[*92*] Sandford, F., "Frost Tests on Bricks," *British Clayworker*, Vol. 68, No. 815, 1960, pp. 18-26.

[*93*] Sandford, F., Lilegren, B., and Jonsson, B., "The Resistance of Bricks Towards Frost—Experiments and Considerations," *Transactions of Chalmers University of Technology*, Gothenburg, Sweden, No. 237, 1961, p. 20.

[*94*] Sandford, F. and Fredholm, H., "Rapid Method for Determining the Frost Resistance of Bricks," *Transactions of Chalmers University of Technology*, Gothenburg, Sweden, No. 327, 1969.

[*95*] Sandford, F., "Super-Cooling of Water in the Pore System of Bricks," *Transaction of Chalmers University of Technology*, Gothenburg, Sweden, No. 330, 1970, p. 11.

[*96*] Sandford, F. and Fredholm, H., "Dependence of the Frost Resistance on the Water-Saturation and the Super-Cooling of Highly Water-Saturated Bricks," *Transactions of Chalmers University of Technology*, Gothenburg, Sweden, No. 328, 1970, p. 19.

[*97*] Stedham, M. E. C., "Rapid Measurement of Water Absorption of Building Bricks," *Transactions of the Eighth International Ceramic Congress*, Copenhagen, 1962, pp. 301-307.

[*98*] Stull, R. T. and Johnson, P. V., "Some Properties of the Pore Structure in Bricks and Their Relation to Frost Action," *Journal of Research*, U.S. National Bureau of Standards, Washington, DC, Vol. 25, 1940; reprinted as Research Paper 1349.

[99] Trelfall, C. R. F., "XXI. Porosity," *Transactions of the British Ceramic Society*, Stoke-on-Trent, England, Vol. 33, No. 8, 1934, pp. 299-320.
[100] Voss, W. C., "Classification of Brick by Water Absorption," *Industrial and Engineering Chemistry*, American Chemical Society, Easton, PA, Vol. 27, 1935, pp. 1021-1022.
[101] Washburn, E. W., "Porosity: I. Purpose of the Investigation, II. Porosity and the Mechanism of Absorption," *Journal of the American Ceramic Society*, Columbus, OH, Vol. 4, 1921, pp. 916-922.
[102] Watson, A., "Surface Growth of Salt Crystals as a Contributory Cause of Frost Damage to Bricks," *Transactions of the British Ceramic Society*, Stoke-on-Trent, England, Vol. 56, No. 7, 1957, pp. 366-368.
[103] Watson, A., "Laboratory Tests and the Durability of Bricks. VI. The Mechanism of Frost Action on Bricks," *Transactions of the British Ceramic Society*, Stoke-on-Trent, England, Vol. 63, No. 11, 1964, pp. 663-680.
[104] Watson, A., "Laboratory Tests and the Durability of Bricks. VII. Frost Dilatrometry as a Routine Test," *Transactions of the British Ceramic Society*, Stoke-on-Trent, England, Vol. 63, No. 11, 1964, pp. 681-695.
[105] Watson, A., "The Mechanism of Frost Action in Bricks," *Transactions of the British Ceramic Society*, Stoke-on-Trent, England, Vol. 63, No. 11, 1964, pp. 663-680.
[106] Whittemore, O. J., "Absorption: Its Relation to Durability," *Journal of the American Ceramic Society*, Columbus, OH, Vol. 13, 1930, p. 80.

Masonry

[107] Symposium on Weathering Characteristics of Masonry Materials, *ASTM Proceedings*, American Society for Testing and Materials, Philadelphia, Vol. 31, No. 31, Part 2, 1931, pp. 715-835.
[108] "Some Aspects of Weathering of Building Materials," *The Builder*, London, England, Vol. 135, 1928, p. 4477.
[109] "The Weathering, Preservation, and Maintenance of Natural Stone Masonry," *Building Research Station Digest*, No. 21, Her Majesty's Stationary Office, London, England, Aug. 1950.
[110] Aufiery, J. T., "Preservation is Intelligent Restoration—Water Versus Preservation," *Procedures of the Second North American Masonry Conference*, University of Maryland, College Park, Md., Aug. 1982, pp. 38.
[111] Bird, E. B. and Allen, W., "Weatherings," *Journal of the Royal Institute of British Architects*, Vol. 47, No. 3, Jan. 1940, pp. 57-61.
[112] Boynton, R. S. and Gutschick, K. A., "Durability of Masonry and Mortar," Masonry Mortar Technical Notes No. 1, National Lime Association, Feb. 1964.
[113] Brady, F. L., "Fundamental Principles of the Weathering of Building Materials," *Journal of the Royal Institute of British Architects*, Vol. 50, No. 8, 1943, pp. 177-180; Architect and Building News, Vol. 174, No. 3876, 1943, pp. 10-11.
[114] Brightly, H. S., "Economic Aspects of Masonry Decay from Weathering," *ASTM Proceedings*, American Society for Testing and Materials, Philadelphia, Vol. 31, Part II, 1931, pp. 716-724.
[115] Christensen, G., "Some Investigations on the Frost Resistance of Insulated Cavity Brick Walls," *Proceedings* of RILEM/CIB Symposium on Moisture Problems, RILEM, Helsinski, 1965, pp. 9.
[116] Cook, J. P., *Construction Sealants and Adhesives*, Wiley-Interscience, New York, N.Y., 1970.
[117] Craske, C. W., "Notes on the Durability of Building Materials," *Surveyor and Municipal and County Engineer*, Vol. 103, No. 2692, 1943, pp. 351-353; No. 2693, pp. 363-364.
[118] Dorey, D. B., "Weather as a Factor in Masonry Problems," *Journal of the Royal Architectural Institute of Canada*, Vol. 32, Oct. 1955; Reprinted by National Research Council of Canada, Division of Building Research, Technical Paper No. 32, NRC 3794, 1955.
[119] Dorsey, N. E., *Properties of Ordinary Water Substance*, Reinhold Publishing Corp., New York, N.Y., 1940.
[120] Foran, M. R., Vaughan, V. E., and Reid, T., "Masonry Deterioration—A Study in the Maritime Provinces," Dalhousie University and Nova Scotia Technical College, 1947, p. 54.

[*121*] Frohnsdorff, G. and Masters, L. W., "The Meaning of Durability and Durability Prediction," *Durability of Building Materials and Components, ASTM STP 691*, American Society for Testing and Materials, Philadelphia, 1980, pp. 17-30.

[*122*] Gutschick, K. A. and Clifton, J. R., "Durability Study of 14-year old *Masonry Wallets*," *Masonry Past and Present, ASTM STP 589*, American Society for Testing and Materials, Philadelphia, 1975, pp. 76-95.

[*123*] Garden, G. K., "Damage to Masonry Constructions by the Ice-Lensing Mechanism," *Proceedings*, RILEM/CIB Symposium on Moisture Problems in Buildings, Helsinki, 1965, Vol. 1, Section 2, Paper 6, 8 pp.

[*124*] Gerard, R., "Standard Frost Resistance Tests in Belgium," *Proceedings*, Fourth International Brick-Masonry Conference, Bruges, April 1976 (in French).

[*125*] Grimm, C. T., "Masonry Maintenance and Restoration—A Guide to the Literature," *Structural Renovation and Rehabilitation of Buildings*, Boston Society of Civil Engineers Section/ASCE, Boston, Mass., Nov. 1979, pp. 71-90.

[*126*] Grimm, C. T., "Coatings for Masonry," *Proceedings* of the Second Canadian Masonry Conference, Carleton University, Ottawa, Ontario, Canada, June 1980, pp. 441-456.

[*127*] Grimm, C. T., "Water Permeance of Masonry Walls: A Review of the Literature," *Masonry: Materials, Properties, and Performance, ASTM STP 778*, American Society for Testing and Materials, Philadelphia, 1982, pp. 178-199.

[*128*] Grimm, C. T., "A Driving Rain Index For Masonry Walls," *Masonry: Materials, Properties, and Performance, ASTM STP 778*, American Society for Testing and Materials, Philadelphia, 1982, pp. 171-177.

[*129*] Gutschick, K. A. and Clifton, J. R., "Durability Study of 14-Year Old Masonry Wallettes," *Masonry: Past and Present, ASTM STP No. 589*, American Society for Testing and Materials, Philadelphia, June 1974, p. 76.

[*130*] Hutcheon, N. B., "Principles Applied to an Insulated Masonry Wall," *Canadian Building Digest*, National Research Council of Canada, Division of Building Research, CBD 50, 1964.

[*131*] Jessop, E. L. and Baker, L. R., *Moisture, Thermal, Elastic, and Creep Properties of Masonry: An Annotated Bibliography*, Center for Research and Development in Masonry, Calgary, Alberta, Canada, Nov. 1980.

[*132*] Kampf, L., "Durability of Masonry," *Journal of Materials*, American Society for Testing and Materials, Philadelphia, Vol. 1, No. 1, 1966, pp. 203-225.

[*133*] Kessler, D. W., "Frost? No, Decay! How to Prevent It," *American Architect*, Vol. 139, No. 2592, 1931, pp. 28-29, 76-78.

[*134*] McBurney, J. W., "Effect of Atmosphere on Masonry and Related Materials," *Symposium on Some Approaches to Durability in Structures, ASTM STP No. 236*, American Society for Testing and Materials, Philadelphia, 1958, pp. 45-56.

[*135*] McIntyre, W. A., *Investigations into the Durability of Architectural Terra Cotta and Faience*, Building Research Special Report No. 12, Her Majesty's Statuary Office, London, 1929.

[*136*] McNeal, R. M., "Effects of Water and Moisture on a Building," *Construction Specifier*, Alexandria, Va., Vol. 27, No. 11, Nov. 1974, pp. 24-29.

[*137*] Palmer, L. A., "Durability and Water-Tightness of Walls of Unit Masonry," *Canadian Engineer*, Vol. 67, No. 2, 1934, pp. 8-9.

[*138*] Plusch, H. A., "Some Causes of The So Called Disintegration of Terra Cotta, Brick, and Allied Ceramic Material and Suggested Remedies," *Ceramic Age*, Newark, NJ, Vol. 1, 1921, p. 65.

[*139*] Ritchie, T. and Plewes, W. G., "Design of Unit Masonry for Weather Resistance," Technical Paper No. 30, National Research Council of Canada, Division of Building Research, NRC 3754, 1955.

[*140*] Ritchie, T. and Davison, J. I., "Moisture Content and Freeze-Thaw Cycles of Masonry Materials," *Journal of Materials*, American Society for Testing and Materials, Philadelphia, Vol. 3, No. 3, Sept. 1968, pp. 658-671; reprinted by National Research Council of Canada, Division of Building Research, Research Paper No. 370, NRC 10297, 1968.

[*141*] Schaffer, R. J., "Some Aspects of Weathering of Building Materials," *The Builder*, London, Vol. 135, 1928, p. 865.

[*142*] Sereda, P. J., "Weather Factors Affecting Corrosions of Metals," *Corrosion in Natural En-*

vironments, ASTM STP 558, American Society for Testing and Materials, Philadelphia, 1974, pp. 7-22.

[*143*] Simpson, J. W. and Horrobin, P. J., *The Weathering and Performance of Building Materials*, Wiley Interscience, New York, 1970.

[*144*] Stambolov, T. and Van Asperen De Boer, J. R. J., *The Deterioration and Conservation of Porous Building Materials in Monuments*, International Centre for Conservation, Rome, 1972, pp. 36-43.

[*145*] Sturrup, V. R., "Exposure Testing of Concrete and Masonry," *Ontario Hydro Research News*, Vol. 11, July-Sept. 1959, pp. 34-37.

[*146*] Thomas, W. N., "Experiments on the Freezing of Building Materials," Building Research Technical Paper No. 17, Department of Scientific and Industrial Research, Great Britain, 1938, pp. 146.

[*147*] Winkler, E. M., *Stone: Properties, Durability in Man's Environment*, Springer-Verlag, Vienna, New York, 1975.

Brick Masonry

[*148*] "Corrosion of Stone and Brick," *Chemical Age*, Morgan-Grantpian, London, Vol. 43, No. 110, 1940, p. 153.

[*149*] "Weathering of Brick Walls," *Engineering News-Record*, Vol. 100, No. 12, 1928, p. 487.

[*150*] Butterworth, B., "The Camerman Theory," *Transactions of the British Ceramic Society*, Stoke-on-Trent, England, Vol. 53, No. 9, 1954, pp. 563-604.

[*151*] Chrisensen, G., "Investigations on the Frost Resistance of Insulated Cavity Brick Walls," *Proceedings*, RILEM/CIB Symposium on Moisture Problems in Buildings, Helsinki, 1965, Vol. 1, Section 3, Paper 12, 9 pp.

[*152*] Churchill, W. M., "Frost Attack of Brickwork—Three Descriptive Case Histories," *Proceedings*, Fourth International Brick-Masonry Conference, Bruges, April 1976.

[*153*] Clews, F. H., "Experiments to Assess the Durability of Bricks and Brickwork," *Proceedings*, British Ceramic Society, Stoke-on-Trent, England, No. 4, July 1965, pp. 93-108.

[*154*] Frank, G. A., "Durability and Protective Measures for Brickwork Exposed to Corrosion by Alkali," (translated from Promishlennost Stroitel'stvo, Vol. 37, No. 4, 1959, pp. 44-46), Department of Scientific and Industrial Research, Great Britain, Building Research Station, Library Communication No. 934, 1960, p. 4.

[*155*] Grimm, C. T. and Halsell, R. D., "Quality Control of Brick Masonry Construction in the U.S.A.," *SIBMAC Proceedings*, British Ceramic Research Assn., Stoke-on-Trent, England, 1971, pp. 337-342.

[*156*] Grimm, C. T., "Thermal Strain in Brick Masonry," *Proceedings of The Second North American Masonry Conference*, University of Maryland, College Park, Md., Aug. 1982, p. 34.

[*157*] Grimm, C. T., in this publication, pp. 67-85.

[*158*] Howard, J. W., Hockaday, R. B., and Soderstrum, W. K., "Effects of Manufacturing and Construction Variables on Durability and Compressive Strength of Brick Masonry," *Designing, Engineering and Constructing with Masonry Products*, Proceedings of The International Conference on Masonry Structural Systems, Austin, TX, 1967, Gulf Publishing Company, Houston, TX, 1969.

[*159*] Johnson, P. V. and Plummer, H. C., "Some Factors Affecting Durability of Structural Clay Products Masonry," *Symposium on Some Approaches to Durability in Structures, ASTM STP No. 236*, American Society for Testing and Materials, Philadelphia, 1958, p. 3.

[*160*] Laurie, A. P., "The Decay of Modern Brickwork," *British Clayworker*, Vol. 35, 1927, p. 280.

[*161*] Lent, L. B., "A Primer on Efflorescence," *The Clay-Worker*, Indianapolis, Ind., Vol. 93, No. 4, March 1930, pp. 280-281.

[*162*] Millstone Grit, "Decay in Brickwork," *The Illustrated Carpenter and Builder*, Vol. 122, No. 3156, 1938, pp. 356, 358.

[*163*] Palmer, L. A. and Hall, J. V., "Durability and Strength of Bond Between Mortar and Brick," *Journal of Research*, U.S. National Bureau of Standards, Vol. 6, No. 3, March 1931, pp. 473-492; reprinted as Research Paper 290.

[164] Palmer, L. A., *The Construction of Weather Resistance Masonry Walls*, Brick Institute of America, McLean, VA, 1936, p. 27.
[165] Ritchie, T., "Damage by Spalling of Painted Brickwork of Former CMHC Houses in Kingston," Technical Note No. 133, National Research Council of Canada, Division of Building Research, 1952.
[166] Ritchie, T., "Frost Action on Brick Work," Technical Note No. 134, National Research Council of Canada, Division of Building Research, 1952.
[167] Ritchie, T., "Relation of Exterior Paints to Surface Decay of Brick Masonry Walls," Internal Report No. 33, National Research Council of Canada, Division of Building Research, 1953.
[168] Vitruvius, *The Ten Books of Architecture*, Dover Publications, New York, 1960.
[169] West, H. W. H., Hodgkinson, H. R., and Davenport, S. T. E., *The Performance of Walls Built of Wire Cut Bricks with and without Perforations*, British Ceramic Research Association, Stoke-on-Trent, England, 1968.

Mortar

[170] Report of the Building Research Board for the Year 1926, Department of Science and Industrial Research, London, 1927.
[171] Boynton, R. S. and Gutschick, K. A., "Bond of Mortar to Masonry Units. Factors Influencing Strength, Extent, and Durability of Bond," Masonry Mortar Technical Notes No. 3, National Lime Association, Washington, DC, Sept. 1964.
[172] Boynton, R. S. and Gutschick, K. A., "Durability of Masonry and Mortar," Masonry Mortar Technical Notes No. 1, National Lime Association, Feb. 1964.
[173] Connor, C. C. and Okerson, W. E., "Recent Disintegration of Mortar in Brick Walls," *ASTM Proceedings*, American Society for Testing and Materials, Philadelphia, Vol. 57, 1957, p. 1170.
[174] Davison, J. I., "Effect of Air Content on Durability of Cement-Live Mortars," *Durability of Building Materials*, Elsevier Scientific Publishing Co., Amsterdam, 1982, pp. 23–34.
[175] Eutache, J. and Magnan, R., "Method of Determining Resistance of Mortars to Sulphate Attack," *Journal of the American Concrete Institute*, Detroit, MI, Vol. 55, No. 5, 1972, pp. 237–239.
[176] Goebel, E., "The Effect of Frost on Lime Putty," (translated from *Zement-Kalk-Gips*, Vol. 6, No. 4, 1963, pp. 252–254), Library Communication No. 544, Department of Scientific and Industrial Research, Great Britain, Building Research Station, 1954, p. 6.
[177] Hughes, C. A. and Anderson, K. A., "Effect of Fine Aggregate on Durability of Mortars," *ASTM Proceedings*, American Society for Testing and Materials, Philadelphia, Vol. 41, 1941, p. 987.
[178] McBurney, J. W., "The Effects of Weathering on Certain Mortars Exposed in Brick Masonry with and without Caps and Flashings," *ASTM Proceedings*, American Society for Testing and Materials, Philadelphia, Vol. 56, 1956, pp. 1273–1287.
[179] Miller, D. G. and Manson, P. W., "Longtime Tests of Concrete and Mortars Exposed to Sulfate Waters," University of Minnesota Agricultural Experiment Station, February 1950, p. 111.
[180] Minnich, L. J., "Effect of Time and Characteristics of Mortar in Masonry Construction," *American Ceramic Society Bulletin*, Columbus, Ohio, Vol. 38, No. 5, 1959, pp. 239–245.
[181] Ritchie, T., "Decay of Masonry Mortar Due to Sulphate Salts," *Internal Report* No. 276, National Research Council of Canada, Division of Building Research, 1963.
[182] Roberts, J. A. and Vivian, H. E., "Further Studies on the Action of Salt Solutions on Cracked Mortar," *Australian Journal of Applied Science*, Vol. 12, No. 3, 1961, pp. 348–355.
[183] Rohland, P., "The Weathering of Stones and Mortars," *Chemical Abstracts*, Vol. 5, 1911, p. 2421.
[184] Scofield, H. H., "Some Tests to Show Effect of Freezing on Permeability, Strength, and Elasticity of Concretes and Mortars," *ASTM Proceedings*, American Society for Testing and Materials, Philadelphia, Vol. 37, Part II, 1937, p. 655.
[185] Sneck, T., Kinnunen, L., and Koski, L., "Investigations on the Properties of Lime-Cement

Mortars at Low Temperatures," RILEM/CIB Symposium on Moisture Problems in Buildings—August 1965, RILEM Bulletin 34, March 1967, pp. 2-32.

[*186*] Wilson, H., "Progress Report on the Efflorescence and Scumming of Mortar Materials," *American Ceramic Society Journal*, Columbus, Ohio, Vol. 11, No. 1, 1928, pp. 1-34.

[*187*] Zemaitis, W. L., "Factors Affecting Performance of Unit-Masonry Mortar," *Journal of The American Concrete Institute*, ACI, Detroit, MI, Dec. 1959, pp. 461-471.

Florescence

[*188*] "Efflorescence," Clay Products Bulletin, New Zealand Pottery and Ceramics Association, Lower Hutt, New Zealand, No. 19, May 1961, p. 1-4.

[*189*] "Efflorescence—Research in Europe and Canada," *Claycraft*, London, Oct. 1963, pp. 2-5.

[*190*] "Efflorescence and Stains on Brick Walls," *Guide to Good Building*, Department of The Environment, Her Majesty's Stationery Office, London, No. 75, 1972.

[*191*] "Soluble Salts in Brickwork," Technical Notes on Clay Bricks, No. 3, Brick Development Research Institute, University of Melbourne, Parkville, Victoria, Australia, Feb. 1976.

[*192*] "Sulphate Attack on Brickwork," *Building Research Establishment Digest*, Building Research Establishment, Her Majesty's Stationery Office, London, 1979.

[*193*] "Efflorescence, Causes," Technical Notes on Brick Construction, No. 23, Brick Institute of America, McLean, Va., July 1981.

[*194*] Anderegg, F. O., "Efflorescence," ASTM Bulletin, American Society for Testing and Materials, Philadelphia, Oct. 1952, pp. 39-43.

[*195*] Bonnell, D. G. R. and Notage, M. E., "Studies in Porous Materials with Special Reference to Building Materials. I. The Crystallisation of Salt in Porous Materials," *Journal of the Society of the Chemical Industry*, London, Vol. 58, Jan. 1939, pp. 16-21.

[*196*] Brady, F. L. and Coleman, E. H., "III—The Effect of Firing Conditions upon the Soluble Salt Content of Clayware," *Transactions of the British Ceramic Society*, Stoke-on-Trent, England, Vol. 30, No. 5, 1931, pp. 169-186.

[*197*] Brady, F. L. and Coleman, E. H., "IV—The Effect of Firing Conditions upon the Soluble Salt Content of Clayware," *Transactions of the British Ceramic Society*, Stoke-on-Trent, England, Vol. 31, No. 2, 1933, pp. 58-73.

[*198*] Brady, F. L. and Butterworth, B., "V—The Staining of Facing Bricks," *Transactions of the British Ceramic Society*, Stoke-on-Trent, England, Vol. 31, No. 6, 1932, pp. 193-201.

[*199*] Brownell, W. E., "Fundamental Factors Influencing Efflorescence of Clay Products," *Journal of the American Ceramic Society*, Columbus, Ohio, Vol. 32, No. 12, Dec. 1, 1949, pp. 379-389.

[*200*] Brownell, W. E., "Application of New Techniques to the Solution of an Efflorescence Problem," *Journal of the American Ceramic Society*, Columbus, Ohio, Vol. 33, No. 12, Dec. 1950, pp. 360-363.

[*201*] Brownell, W. E., "Efflorescence Resulting from Sulfates in Clay Raw Materials," *Journal of the American Ceramic Society*," Columbus, Ohio, Vol. 41, No. 8, August 1958, pp. 310-314.

[*202*] Brownell, W. E., "Retention of Sulfates by Fired Clay Products," *Journal of the American Ceramic Society*, Columbus, Ohio, Vol. 43, No. 4, April 1960, pp. 179-183.

[*203*] Brownell, W. E., *The Causes and Control of Efflorescence on Brickwork*, Research Report No. 15, Brick Institute of America, McLean, VA, August 1969.

[*204*] Butterworth, B., "VI. The Florescence Test and the Chemical Examination of Florescence," *Transactions of the British Ceramic Society*, Stoke-on-Trent, England, Vol. 36, No. 6, 1933.

[*205*] Butterworth, B., "VII. The Relation of Soluable Salt Content to Florescence," *Transactions of the British Ceramic Society*, Stoke-on-Trent, England, Vol. 35, No. 3, 1936, pp. 105-118.

[*206*] Butterworth, B., "The Practical Significance of Tests for Soluable Salts in Clay Products," *Transactions of the British Ceramic Society*, Stoke-on-Trent, England, Vol. 52, No. 4, 1953, pp. 174-178.

[*207*] Cooling, L. F., "Contributions to the Study of Florescence," *Transactions of the British Ceramic Society*, Stoke-on-Trent, England, Vol. 29, No. 2, 1930, pp. 39-52.

[*208*] Cooling, L. F., "II. The Evaporation of Water from Bricks," *Transactions of The British Ceramic Society*, Stoke-on-Trent, England, Vol. 29, 1929, pp. 34-54.

[*209*] Davidson, J. I., A Study of Efflorescence in Clay Bricks," *Canadian Ceramic Society Journal*, Toronto, Ontario, Vol. 35, 1966, pp. 85-92.

[*210*] Goodwin, M. J., "Efflorescence on the Exterior Surface of Masonry Walls," Building Note No. 8, National Research Council of Canada, Ottawa, Canada, July 1950.

[*211*] Hardesty, J. M., "Disintegration of Face Bricks by Dissolved Salts," *Bell Laboratories Record*, Vol. 22, No. 5, Jan. 1944, pp. 222-224.

[*212*] Jackson, F. G., "A Descriptive Bibliography of Scumming and Efflorescence," *Bulletin of the American Ceramic Society*, Columbus, Ohio, Vol. No. 4-8, 1925, pp. 376-401.

[*213*] Kampf, L. and Rogers, P. L., "Discussion of a Paper on a Method of Test for Potential Efflorescence of Masonry Mortar," ASTM Bulletin, American Society for Testing and Materials, Philadelphia, Sept. 1959, pp. 45-46.

[*214*] Kass, I., "Controlling Efflorescence on Building Brick," *Ceramic Age*, Newark, NJ, Vol. 18, No. 2, Aug. 1931, pp. 85-86.

[*215*] Laurie, A. P. and Milne, J., "Evaporation of Water and Salt Solutions from Surfaces of Stone, Brick and Mortar," *Proceedings of the Royal Society of Edinburgh*, Edinburgh, Scotland, Vol. 47, No. 4, 1926-27, pp. 52-68.

[*216*] Lim, B. S. and Cutler, I. B., *Efflorescence of Masonry Mortars*, Department of Materials Science and Engineering, University of Utah, Salt Lake City, Utah, Jan. 1976.

[*217*] McIntyre, W. A., "Efflorescence on Brickwork," *The British Clayworker*, Watford, Herts, England, Oct. 1978, pp. 252-254.

[*218*] Palmer, L. A., "Cause and Prevention of Kiln and Dry-House Scum and of Efflorescence on Face-Brick Walls," *Technologic Papers of the Bureau of Standards*, No. 370, Superintendent of Documents, U.S. Government Printing Office, Washington, DC, Vol. 22, 1928, pp. 579-629.

[*219*] McBurney, J. W. and Parsons, D. E., "Wick Test for Efflorescence of Building Brick," RP 1015, *Journal of Research of the National Bureau of Standards*, Vol. 19, July 1937.

[*220*] Redfern, C. A., "Efflorescence on Brickwork: Its Causes and Cure," *Architecture*, Vol. 52, 1925, pp. 360-361.

[*221*] Ritchie, T., "Study of Efflorescence on Experimental Brickwork Piers," *Journal of the American Ceramic Society*, Columbus, Ohio, Vol. 38, No. 10, Oct. 1955, pp. 357-361.

[*222*] Ritchie, T., "Study of Efflorescence Produced on Ceramic Wicks by Masonry Mortars," *Journal of the American Ceramic Society*, Columbus, Ohio, Vol. 38, No. 10, Oct. 1955, pp. 362-366.

[*223*] Ritchie, T., "Efflorescence," *Canadian Building Digest*, National Research Council, O'Hara, Ontario, Canada, No. 2, Feb. 1960.

[*224*] Ritchie, T., "The Influence of Efflorescence on Decay and Expansion of Mortar," *Canadian Ceramic Society Journal*, Toronto, Ontario, Canada, Vol. 35, 1966, pp. 92-95.

[*225*] Ritchie, T., "Efflorescence on Masonry," *Moisture Problems in Buildings*, RILEM/CIB Symposium, Helsinki, 1965, Technical Paper No. 223, National Research Council, Ottawa, Canada, July 1966.

[*226*] Robertson, R. F., "Efflorescence," *Canadian Ceramic Society Journal*, Toronto, Ontario, Canada, Vol. 13, No. 40, 1944, pp. 40-43.

[*227*] Rogers, P. L., "A Method of Test for Potential Efflorescence of Masonry Mortars," ASTM Bulletin, American Society for Testing and Materials, Philadelphia, Jan. 1959, pp. 31-33.

[*228*] Young, J. E., "Backup Materials as a Source of Efflorescence," *Journal of the American Ceramic Society*, Columbus, Ohio, Vol. 40, No. 7, July 1, 1957, p. 240.

Summary

This STP is a compilation of papers dealing with the performance of masonry. Generally, the first papers in the volume deal with mortar materials and properties. The remaining papers deal with masonry units, assemblages, and applications.

The first paper presents a discussion by Isberner of methods available to determine the composition of hardened masonry mortar. Data gathered from petrographic examination of mortar can be combined with computer manipulation of quantitative chemical analytical data to yield approximate original composition of hardened mortars. Careful sampling of the mortar is important, and more accurate results can be attained if samples of the component materials of the mortar are also analyzed.

Gazzola et al describes the results of flexural bond strength testing using the bond wrench. The effects of different units, mortars, and sands on the bond strength between mortars and units are presented as is a comparison of different bond wrench arm lengths and testing conditions.

Isberner's second paper presents a method of testing mortars for efflorescence. Some other methods are discussed. The author's method allows a comparison of efflorescing tendencies of different mortar materials. It is possible to collect and analyze the efflorescing salts at different depths in the mortar. The author points out that even materials showing little or no efflorescence in the test could effloresce in actual service.

The paper by Matthys and Chanprichar discusses the ultimate strength flexural theory for reinforced brick masonry. The authors tested 20 beam column specimens and corresponding prisms with high-strength and low-strength brick masonry and with compressive loading parellel to and perpendicular to the bed joints. Flexural compressive stress distribution was determined.

Grimm, in his paper on Corrosion of Steel in Brick Masonry, explains that water and other substances corrode metal connectors, reinforcement, and structural supports. Corrosive conditions vary widely from place to place, depending on humidity, pressure of water in the wall, pollutants, and contaminants. Water can come in contact with metal through penetration and condensation. Corrosion can be very severe in a relatively short period of time. Recommendations for slowing down corrosion are presented by the author.

Harris presents a method of efflorescence testing similar in principle to the one presented on p. 29 by Isberner except that both water permeance and efflorescence are measured by this method. The amount of water penetrating a small mortar specimen is measured at desired increments of time (in days).

When permeability testing is finished, efflorescence is measured by quantitative chemical analysis of the salts deposited on the surface of the mortar specimens.

Yi and Carrasquillo report the results of a research project on the development of a method of test to determine the coefficient of thermal expansion of brick that have been placed in service. The research deals with the effect, on the thermal behavior of brick, of factors such as the moisture content of the brick, the temperature range, the number of temperature cycles, and the location on the brick of the measuring instrumentation. Both new brick, that is, brick that had not been placed in service, and brick that had been in service for at least ten years, were studied. Based on the test results, a proposed method of test using electrical resistance strain gages was developed.

Carr et al investigated the common method of providing temporary support for masonry walls during construction that consists of inclined braces restrained at the ground with stakes. Research was conducted on the stakes capacity to resist horizontal loads. Two sizes of wooden stakes and one size of steel stakes, each at two embedments and three inclinations were tested. Three different soils at three consistencies were used. The authors concluded that for most soils the stake capacity was found to be insufficient to support tall masonry walls in modest wind with the bracing at a reasonable spacing. An alternate method of bracing, using a wooden truss to deliver vertical loads to the soil rather than horizontal loads is suggested.

Naish reports on an effort to analyze the durability performance of glazed brick by conducting a historical review of the properties of glazed brick produced by his company and the characteristics of buildings constructed with glazed brick produced by his company. Mr. Naish concludes that the wall failures that he has observed are not the results of too generous ASTM specifications or inconsistant brick quality but rather wall design, construction, or maintenance. The author suggests two approaches to future research to develop wall design, workmanship, and maintenance criteria to improve the performance of masonry walls.

Huizer and Ward have previously reported on the development of a clay unit to compete the metal chimney. This clay unit has been shown to pass the requirements for "zero clearance" application at hearth temperatures of approximately 700°C (1292°F). Anticipating the probability that this clay unit would be required to meet the more severe requirements of the metal units, the authors investigated four of the tests that were considered most relevant to clay chimneys from the standards for metal chimneys. These tests included thermal shock, 650°C flue gases, 925°C flue gases, and creosote burnout. The test results showed that at present the clay unit would not pass the new standard for metal chimneys. The authors suggest that further research with regard to the unit geometry and the clay mixtures is needed.

Hamid et al report the results of their study to evaluate the use of direct modeling of ungrouted and grouted concrete block masonry under axial compression. The study includes the effects of mortar and grout strength along with various high to thickness ratios, number of courses, and bonding pattern. The degree of correlation with the prototype properties varied; however, the authors conclude that direct modeling is feasible for predicting the behavior of block masonry.

The paper by Wong and Drysdale, like the previous paper, deals with the compression characteristics of concrete block masonry assemblages. Concrete block prisms of various heights, including hollow, solid and grout-filled, were tested for compression normal to the bed joint and parallel to the bed joint. Test results with regard to failure modes, compressive strengths, and stress-strain relationships are discussed. Conclusions and recommendations include comments regarding unit testing, prism construction, wall construction, and design standards.

Johnson points out that there are many differing opinions regarding the appropriate method for the cleaning of masonry. He suggests that there has been a continual search for a universal technique despite the fact that masonry is highly variable. To address this problem, the author presents a strategy for the selection of a cleaning method that utilizes a decision chart to consider these variables. He requests constructive criticism of the decision chart.

Rath's paper addresses the nonperformance of exterior nonload bearing walls, curtain walls, built of brick masonry. He notes that these walls can be constructed in a number of ways, but regardless of the wall type, their function is to provide a permanent barrier to protect the building interior from the outside elements. The categories of nonperformance discussed by the author are water penetration, durability, efflorescence and structural behavior (cracking). He suggests the primary factors effecting nonperformance are design, construction, materials, and ASTM specifications. The author reviews the methods for evaluating and suggests ways of preventing nonperformance.

Grimm's review of the literature on the durability of brick masonry summarizes the engineering literature on this subject published in English since 1900. Bibliographies with a total of 228 entries are provided. It is noted that durability is a function of materials, design, construction environment, and maintenance. It may also be defined as the ability of a material to remain serviceable with prudent maintenance during a normal life span in the intended environment. Agents and mechanics of destruction, porosity, freeze-thaw resistance, mortar properties, florescence, environment architectural engineering, brick specifications, construction, and maintenance are discussed. The conclusions suggest that keeping the masonry as dry as possible is the single most important variable in masonry durability.

As was reported in the introduction to this publication, the effort to encourage the submission of papers dealing with field application and end use problems was successful. The papers addressing these areas range from guidelines for temporary masonry wall support systems and a stratgey for masonry cleaning through corrosion of steel in masonry and end use performance of masonry and masonry materials. The other papers contained herein report on research and test methods that address field application and field problems. Hopefully, in addition to providing a better understanding of the performance of masonry, these papers will generate further research.

John T. Conway

Santee Cement Company, Holly Hill, SC 29059; Symposium cochairman and editor

John C. Grogan

Brick Institute of America Region Nine, 8601 Dunwoody Place, Suite 507, Atlanta, GA 30338; symposium cochairman and editor.

Index

C

D

E

F

G

H

Q

R

S

T

U

V

W

Y